Materials Education

MATERIALS RESEARCH SOCIETY
SYMPOSIUM PROCEEDINGS VOLUME 1233

Materials Education

Symposium held December 1–3, 2009, Boston, Massachusetts, U.S.A.

EDITORS:

M.M. Patterson

University of Wisconsin-Stout
Menomonie, Wisconsin, U.S.A.

E.D. Marshall

New York Hall of Science
Queens, New York, U.S.A.

C.G. Wade

IBM Almaden Research Center
San Jose, California, U.S.A.

J.A. Nucci

Cornell University
Ithaca, New York, U.S.A.

D.J. Dunham

University of Wisconsin-Eau Claire
Eau Claire, Wisconsin, U.S.A.

Materials Research Society
Warrendale, Pennsylvania

CAMBRIDGE UNIVERSITY PRESS
Cambridge, New York, Melbourne, Madrid, Cape Town,
Singapore, São Paulo, Delhi, Mexico City

Cambridge University Press
32 Avenue of the Americas, New York NY 10013-2473, USA

Published in the United States of America by Cambridge University Press, New York

www.cambridge.org
Information on this title: www.cambridge.org/9781107408029

Materials Research Society
506 Keystone Drive, Warrendale, PA 15086
http://www.mrs.org

© Materials Research Society 2010

First published 2010
First paperback edition 2012

Single article reprints from this publication are available through
University Microfilms Inc., 300 North Zeeb Road, Ann Arbor, MI 48106

CODEN: MRSPDH

ISBN 978-1-107-40802-9 Paperback

This material is based upon work supported in part by the National Science Foundation,
Division of Materials Research, under Grant Number DMR-0951410. Any opinions,
findings, and conclusions or recommendations expressed in this material are those of
the author(s) and do not necessarily reflect the views of the National Science Foundation.

CONTENTS

*Invited Paper

PREFACE

Increased focus on educational transformation in STEM disciplines has led to a burgeoning field of materials education research. Its style is largely based on physics education research (PER), but some contributors also model the scholarship of teaching and learning (SoTL). Materials Education's scope is growing rapidly. This volume captures the events of Symposium PP, "Materials Education," held December 1–3 at the 2009 MRS Fall Meeting in Boston, Massachusetts. An inspirational story motivated the unique focus of this symposium: accessible education at all levels, "K through Grey."

Marni Goldman was honored as a pioneer in the field, promoting excellence and accessibility in materials research education at Stanford, while also promoting educational opportunities for the disabled. The work presented in this volume therefore spans traditional materials education research in higher education; SoTL work in K-12 schools, higher education, museums and outreach organizations; accessibility research for learners and professionals with disabilities (ranging from acute physical to "hidden" mental); and materials education and outreach programs of Materials Research Science and Engineering Centers (MRSECs), Nanoscale Science and Engineering Centers (NSECs) and the Nanoscale Informal Science Education Network (NISE Net).

This symposium would not have been able to coalesce, nor would speakers and participants be able to focus on cross-cutting issues affecting different audiences, without the tireless efforts of Materials Research Society staff and generous financial support from the National Science Foundation (DMR-0951410), the IBM Almaden Research Center, Hysitron, Stanford University, and the Goldman Family.

M.M. Patterson
E.D. Marshall
C.G. Wade
J.A. Nucci
D.J. Dunham

March 2010

ix

MATERIALS RESEARCH SOCIETY SYMPOSIUM PROCEEDINGS

MATERIALS RESEARCH SOCIETY SYMPOSIUM PROCEEDINGS

Prior Materials Research Society Symposium Proceedings available by contacting Materials Research Society

Mater. Res. Soc. Symp. Proc. Vol. 1233 © 2010 Materials Research Society   1233-PP04-27

Teaching materials' properties to K-12 students using a sensor board

Theodoros Pierratos[1], Evangelos Koltsakis[1] and Hariton M. Polatoglou[1]
[1]Physics Department, Aristotle University of Thessaloniki, 56124 Thessaloniki, Greece

ABSTRACT

In this work we present two teaching modules, based on the combination of Scratchboard and Scratch, to be used for the study of thermal properties of materials, such as thermal conductivity and heat capacity. These properties are very important for the understanding of many applications. In the design of the modules we have taken into account two scenarios, one for elementary and secondary school students and one for high school students. This determines not only the type of measurement and the analysis of the data, but also the Scratch interface. The main emphasis for the lower grades is placed on the introduction of the concepts and a demonstration of the differences of the properties of different materials, while for the upper grades for making accurate measurements through inquiry based projects. Both modules have been implemented in a high school laboratory, providing reliable measurements and engaging the students in a higher level than usually.

INTRODUCTION

The progress of ICT (Information Communication Technologies) provides tools for improving the quality and the ease of access to education and training. For science teaching, data acquisition systems are a powerful tool for real-time measurements and the presentation of various physical quantities in school science laboratories. These systems are generally quite expensive and sophisticated for the average students' use. In this work an alternative system is proposed: the combination of Scratchboard and Scratch. Both of them have been developed in the MIT Media Labs; Scratch Board is a sensor board coming with several embedded sensors. On the other hand, Scratch is a new freeware multilingual programming environment that can be easily used by pupils to create their own animated stories, video games and interactive art and share their creations with each other across the Internet [1-4]. Both Scratch and Scratchboard have been developed to be used in a very wide field of activities. This system is not specifically addressed to science lessons' activities; there are integrated data acquisition systems for such a use available. One nice feature of Scratchboard is that it has several sensors embedded and comes to a very low price, less than fifty Euros. In addition, it communicates with Scratch, which is a LOGO-based education oriented software. This gives rise to the research question: can this system be used to study thermal properties of materials such as thermal conductivity and heat capacity in school laboratories? Seeking an answer to this question, we designed two teaching modules and we are also in the process of preparing a new set of activities on subjects like optical and electrical properties.

Theory

The steady state conduction is the form of conduction which happens when the temperature difference is constant, so that after an equilibration time, the spatial distribution of temperatures in an object does not change (for example, a bar may be cold at one end and hot at the other, but the temperature gradient along the bar does not change with time). In short, temperature at a section remains constant and it varies linearly along the direction of heat transfer [5]. In the steady state conduction, the amount of heat entering a section is equal to the amount of heat coming out. There also exist situations where the temperature drop or raise occurs more drastically and the interest is in analyzing the spatial change of temperature in the object over time. This mode of heat conduction can be referred to as *unsteady mode of conduction* or *transient conduction* [6]. Analysis of these systems is more complex and (except for simple shapes) calls for the application of approximation theories and/or numerical analysis with the use of a computer [7]. The mathematical description of transient heat conduction yields a second-order parabolic partial-differential equation [8]:

$$\frac{\partial U}{\partial t} = \kappa \nabla^2 U \tag{1}$$

where κ is the thermal diffusivity and $U(x,t)$ is the temperature.

When applied to regular geometries such as infinite cylinders, spheres, and planar walls of small thickness, the equation is simplified to one having a single spatial dimension. With specification of an initial condition and two boundary conditions, the equation can be solved using separation of variables -- leading to an analytical expression for temperature distribution in the form of an infinite series. The time-honored Heisler charts were generated a half-century ago using a one-term approximation to the series, and have been used widely ever since for 1-D, transient-conduction applications [9]. The general solution of equation (1) is [8]:

$$U(x,t) = e^{-\kappa t/\lambda^2} \left[Dcos\left(\frac{x}{\lambda}\right) + Esin\left(\frac{x}{\lambda}\right) \right] \tag{2}$$

where

$$\frac{1}{\lambda^2} = -\frac{1}{\kappa T}\frac{dT}{dt} \tag{3}$$

EXPERIMENT

Scratchboard is connected to a computer through a serial-to-USB cable that comes together with the board. The communication protocol is the RS-232 and the transfer baud rate is 38.4k. The connection between Scratch and Scratchboard is realized using sensing blocks. Sensor value blocks give readings ranging from 0 to 100 or ranging from 0 to 1023 if the user prefers to read raw 10-bit data [10]. These values can be used to control graphic effects or can be stored in a log file.

In a stanchion we placed a steel bar with 30 cm length and 5mm diameter (Figure 1a).

2

Figure 1. a. The steel bar is heated and three thermistors record its temperature at three different points. **b.** Two horseshoe aluminum bars with a thickness of 3mm and 8mm respectively are heated and two thermistors record the temperature of each one at some point. **c.** Two horseshoe bars (a copper one and an aluminum) with a thickness of 3mm are heated and two thermistors record the temperature of each one at some point.

Alternatively we used horseshoe bars of the same material (aluminum) and different thickness (3mm – 8mm) (Figure 2b) or the same thickness (3mm) and different material (copper – aluminum) (Figure 2c). In all cases one of the ends of the bar was placed above a candle flame. The rate of combustion should remain constant for all the experiments in order to provide heat at a constant rate. Along the bars thermistors were placed, connected with Scratchboard through alligator clips. As the bars were heated, the resistance of the thermistors was raised and their values were recorded. As these values are in arbitrary units; we used a digital thermometer for calibration. According to the level of the students we addressed the Scratch's interface differently. For elementary level students a bar was shown, with its color changing from point to point according to the chance of the temperature. For secondary level students graphs were provided that presented the change of temperature versus time (Figure 2).

Figure 2. A Scratch's screenshot.

DISCUSSION

Students realize the setting presented in Fig. 1a and start the measurements. Data collected for about 40 minutes are presented in Fig. 3.

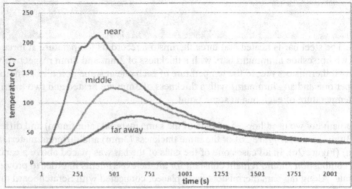

Figure 3. Temperature vs time at three points along a steel bar.

The heating of the bar's end is terminated at the moment that the temperature of the nearest sensor reaches its maximum (blue line). It is obvious in Fig. 3 that the temperature acquired at various points of the bar depends on their distance from the heating source. The closer to the source they are, the higher their temperature. It is also obvious that, for bar points far from the heat source, the temperature continues to increase, even after the interruption of heating (energy offer), due to the heat conduction from the hotter end to the cooler end of the bar. Finally, during the cooling process, curves are compatible to Newton's law of cooling. Students then realize the setting seen in Fig. 1b. Data collected for 40 minutes are presented in Fig. 4.

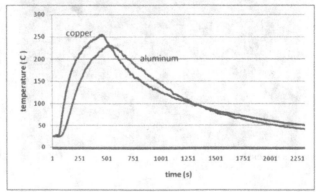

Figure 4. Temperature vs time at two points of a copper bar and an aluminum bar.

4

From Fig. 4, we can observe that the copper bar reaches faster higher temperatures than the aluminum bar, because of its higher thermal conductivity and lower heat capacity. Finally, students constructed the setting presented in Fig. 1c and started the measurements. Data collected for about 15 minutes are presented in Fig. 5.

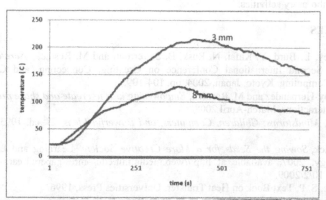

Figure 5. Temperature vs time at one points of two aluminum bars with different thickness.

Two phenomena can be distinguished in Fig. 5: a) because the thicker bar has a larger thermal conductivity, its temperatures increases more rapidly (at least in the beginning), and b) due to its bigger mass, the thicker bar's heat capacity is larger than the thinner bar's, resulting in a higher final temperature for the thinner bar. The graphic representations that came out are in very good agreement with results of simulations [11], [12], and they provide the chance to discuss the laws about the phenomena through inquiry based hands on activities.

It is quite obvious that this setup can provide an accurate measurement of the physical quantities involved after the thermistors' calibration. In addition the low cost apparatus offers the chance to all students in a science class to study a law of Physics by participating in such a hands-on activity. At least in Greek schools, students usually just watch their teacher performing a high accuracy measurement experiment, using an expensive data acquisition system.

CONCLUSIONS

Data acquisition systems provide a powerful tool for measuring and presenting online various physical quantities in school science laboratories. As these systems are in general quite expensive and sophisticated, the use of Scratchboard and Scratch could provide an alternative solution for most schools' laboratories. According to our experience, the appropriate use of such a system amplifies the possibility for students to be engaged in the learning process through programming, inquiry based and hands-on activities. Technically speaking, the sensor board can measure several physical quantities simultaneously and the powerful user-friendly software makes the presentation of the data as well as their processing easy. There is no doubt that the proposed system cannot totally replace the specialized data acquisition systems in upper secondary schools, where the students do have to make precise measurements. But it could be used in a very promising way in lower secondary and in primary education, where sophisticated data acquisition systems usually do not exist. This way would alter students' interest about science and laboratory activities.

REFERENCES

1. J. Maloney, L. Burd, Y. Kafai, N. Rusk, B. Silverman and M. Resnick, *Scratch: A Sneak Preview*. Second International Conference on Creating, Connecting, and Collaborating through Computing. Kyoto, Japan, 2004, pp. 104-109.
2. A. Monroy-Hernández and M. Resnick, *Empowering kids to create and share programmable media*. Interactions, March-April 2008.
3. S. Papert, *Mindstorms: Children, Computers, and Powerful Ideas*, 2nd ed. 1993, NY: Basic Books.
4. M. Resnick, *Sowing the Seeds for a More Creative Society*. Learning and Leading with Technology, 2007, available at http://web.media.mit.edu/~mres/papers/Learning-Leading-final.pdf, 10/2/2009.
5. Sukhatme, S. P, Text Book on Heat Transfer, Universities Press, 1996.
6. Introduction to Transient Conduction and Convection, http://www.ni.com/pdf/academic/us/me105_lab2_2003.pdf
7. Transient Conduction, Retrieved from http://www.philonnet.gr/downloads/ansys/flowlab/unsteady_tutorial.pdf, 1/8/2009.
8. Weisstein, Eric W. "Heat Conduction Equation." From *MathWorld*--A Wolfram Web Resource. http://mathworld.wolfram.com/HeatConductionEquation.html
9. R.J. and O'Leary, G.W., "A Teaching Module for One-Dimensional, Transient Conduction", *Computer Applications in Engineering Education*, Vol. 6, pp. 41-51, 1998.
10. Scratch web site http://info.scratch.mit.edu/@api/deki/files/1162/=ScratchBoard_Tech_InfoR1_(1).pdf
11. L. G. Leal (1992) Laminar flow and Convective Transport Processes, Butterworth pp 139-144.
12. Mark J. McCready (1998) Solution of the Heat Equation for transient conduction by Laplace Transform, Retrieved from http://www.nd.edu/~mjm/heatlaplace.pdf, 10/9/2009.

Mater. Res. Soc. Symp. Proc. Vol. 1233 © 2010 Materials Research Society   1233-PP01-07

Deaf and Hard of Hearing Undergraduate Interns Investigate Smart Polymeric Materials

Peggy Cebe[1], Dan Cherdack[1], Robert Guertin[1], Terry Haas[2], Wenwen Huang[1], B. Seyhan Ince-Gunduz[1], Roger Tobin[1], Regina Valluzzi[1]

[1]Physics and Astronomy Department, Tufts University, Medford, MA 02155, U.S.A.
[2]Chemistry Department, Tufts University, Medford, MA 02155, U.S.A.

ABSTRACT

Smart Materials are those which can undergo a reversible property change in response to an external influence. An important polymeric Smart Material is poly(vinylidene fluoride), or PVF_2, which is piezoelectric. The structure of PVF_2 determines which crystal phases will be electrically active. Recent research has shown that the electrically active beta phase of PVF_2 grows preferentially in nanocomposites of PVF_2 mixed with organically modified silicates (OMS) [1-4]. These nanocomposite Smart Materials offer a new processing strategy for PVF_2 piezo-films. Using PVF_2 nanocomposites as the research focal point, a summer internship program was developed for deaf and hard of hearing (DHH) undergraduate students [5,6]. This paper describes the program and presents research results achieved by the interns. It is written from the perspective of the Principal Investigator, Cebe, on behalf of all the co-authors.

INTRODUCTION

Beginning in summer of 2003, an internship program for DHH students was developed by one of the authors (PC) based on discussions and contacts with the Department of Chemistry and Physics at Gallaudet University in Washington, DC. Gallaudet is the Nation's premiere Liberal Arts University whose mission is specifically to educate deaf and hard of hearing students. Early in that year, I (author PC) had presented a lecture at Gallaudet titled "What Superman Sees with X-ray Vision." Given the interest and excitement of the audience, I inquired whether it would be possible to have a DHH student perform summer research in my laboratory at Tufts University. The enthusiastic reception of this idea by the Gallaudet science faculty suggested there is a strong need for summer internships in the hard sciences (i.e., Science, Technology, Engineering, and Mathematics, or STEM disciplines) for DHH students.

With funding from the National Science Foundation, I began a pilot program with a single intern in summer of 2003. This first summer allowed me to work out the communication difficulties that might be encountered by a deaf student working in a "hearing" environment. In preparation, I learned some basic words in American Sign Language, and as well, I taught myself slowly to finger spell the letters of the alphabet. I still recall the joy I felt the first time I signed to the intern "Please come" from a distance of 40 feet, and she immediately approached me.

The members of my research group, graduate and undergraduate students, welcomed the addition of the Gallaudet student, who spent eight weeks performing research on thin films. This first summer served to build my own confidence that I could, with help from my colleagues and the students in my research group, provide a scientifically rigorous learning experience within a "hearing" environment for DHH students, and communicate effectively by using writing tablets or by typing on a laptop computer.

The longer range goals, to be implemented after the pilot program, were: to bring a small team of four or five DHH students to Tufts for a 6 week summer internship titled "Polymer-based Nanocomposites;" to use a Materials Science and Engineering approach to elucidate the processing-structure-property relationships; to provide an integrated classroom and laboratory learning experience; and to provide intensive guidance and mentoring for DHH students to bring them more fully into scientific research. Classroom and laboratory components were reviewed previously [5,6]. In this paper, I present additional aspects of the recruitment, demographics, communication, and training, and recent results of the interns' investigation of Smart Materials.

RESULTS AND DISCUSSION

Intern Recruitment

Interns are recruited primarily from Gallaudet University (GU) and Rochester Institute of Technology (RIT), which is home to the National Technical Institute for the Deaf (NTID). I work closely with the science and engineering departments and with the employment coordinators to attract high quality applicants to the internship. A flier announcing the internship details is sent to contacts at GU and RIT, as well as being available on my groups' website (http://ase.tufts.edu/~cebe). Applicants submit a resume/curriculum vitae, statement of purpose, unofficial transcript, and letter of recommendation from an advisor or teacher. Successful applicants must have at least a B-grade point average (3.0 out of 4.0), be majoring in any area of science, engineering or technology, and be viewed as able to work well in a team environment. The National Science Foundation supports this program and provides the interns a stipend, housing in Tufts' dormitory, food allowance, and tuition and fees for a 1/2 credit 6-week long summer course taught by me and co-authors Guertin, Hass, Tobin, and Valluzzi. American Sign Language interpreters are provided for the course-work portion of the internship. Because this is a very labor intensive internship, salary support for a laboratory teaching assistant is also provided (in different summers co-authors Cherdack, Ince-Gunduz, or Huang were the TA's).

Intern Demographics

Through summer of 2009, twenty six DHH interns participated in this program. **Figure 1** shows the demographics of the interns, according to disability status, sex, race/ethnicity, and home institution. All interns were deaf or hard of hearing, and eight had American Sign Language (ASL) as their primary language ahead of English. These eight students did not voice, hear, or lip-read. For these students, communication in the laboratory setting was exclusively via written means, such as writing tablets or laptop computers.

The percentage of female interns (58%) and black/Hispanic interns (23%) far exceeds what would be found STEM disciplines generally [7, 8]. There is strong overlap between some categories in the chart. For example, six of the students listed in the category ASL (those for whom ASL was first language; these students did not hear or voice) were females, and four were black. Thus, the most disadvantaged student participants, in terms of severity of hearing disability, happen also largely to come from under-represented demographic groups. We have met one of the major goals of this program, to provide an opportunity and increase participation in research for minority populations including women, black/Hispanic, and disabled students.

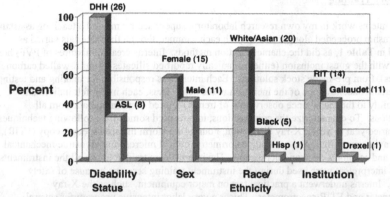

Figure 1. Demographics of 26 interns from 2003-2009, according to the categories: Disability Status, Sex, Race/Ethnicity, and Institution. Tag lines show the number of interns out of 26.

Communication in the Laboratory

In one summer, three of four interns did not voice. To ensure effective communication among all members of the group, we had meetings using an interactive "chat room", provided through Tufts "Blackboard" academic web site. **Figure 2** shows four interns and two instructors all using laptops, and communicating in real time through the chat room. The interns enjoyed this format of group meeting, since it had both a high-tech and a social aspect, and allowed all

Figure 2. Instructors and interns are shown using their laptop computers to access a live Chat Room. This facilitated group meetings for the interns who did not voice.

students to participate equally. One disconcerting aspect of the interactive chat format, and one to which I had to become accustomed, is the non-linear nature of the interaction: multiple conversations are going on at once, with the interns joking back and forth, asking and answering each others' queries in a jumbled sequence.

Laboratory Training

The interns work in my own research laboratory, supervised by me and a teaching assistant. The internship proceeded along a similar path each summer, though the materials varied as indicated in **Table 1**, as did the characterization methods. Interns created mixtures of PVF_2 host polymer with the guest inclusion (either organically modified silicates or multi-walled carbon nanotubes), from pre-made stock solutions. Each intern was responsible for making and testing two different concentrations of the inclusion in the host. Thus, each intern felt individual responsibility to the team, since observation of trends required combining data from all compositions. To characterize the compositions, interns used some of the following techniques (which varied year to year): X-ray diffraction, Fourier transform infrared spectroscopy (FTIR), thermogravimetry, differential scanning calorimetry, optical microscopy, dynamic mechanical analysis, and scanning electron microscopy. The interns were trained to use all the instruments, and ASL interpreters were used during the instrument training sessions because of safety concerns. Interns underwent a practical test on major equipment, such as the X-ray diffractometer and FTIR spectrometer. This is a very labor intensive program for several reasons. First, the students are novices and need a certain level of hand holding in the first three weeks to guarantee their safety; second, daily assessment and feedback are needed to keep forward progress. The interns rose to the challenge, and after an initial period it became possible for them to organize and plan their own work.

Table 1. The guest inclusion and thermal treatment varied by year and lead to different crystal phase of the host polymer, PVF_2 *.

Year	Inclusion	Treatment	PVF_2 Crystal Phase
2004	Lucentite OMS [a]	145 °C [d]	Beta favored [3]
2005	Lucentite OMS	150 °C [e]	Gamma favored [9]
2006	Lucentite OMS	164 °C [e]	Gamma favored
2007	Cloisite OMS [b]	150 °C [e]	Alpha favored
2008	Carbon Nanotubes	Quenched [f]	Alpha favored [10]
2008	Carbon Nanotubes [c]	ZD 105 °C [g]	Beta favored [10]

    PVF_2 Kynar resin, grade 740 pellets were obtained from Elf Atochem.
[a]   Lucentite STN powder was obtained from Zen-Noh Unico, Americas.
[b]   Cloisite 20A powder was obtained from Southern Clay Products, Inc.
[c]   Multi-walled CNTs were purchased from MER Corporation.
[d]   Crystallization was from quenched glassy state.
[e]   Crystallization was from the molten state.
[f]   Samples were compression molded at 200 °C and air quenched.
[g]   Samples were oriented by zone drawing at 105°C.

After passing the practical equipment exam, the interns were qualified to start up the X-ray generator from complete shut down, and run their samples unaided. By week four it becomes necessary to divide the jobs so that each intern would "specialize" in one or two techniques. The project goal is well defined: all eight compositions must be tested by all methods of characterization, the data analyzed, and trends plotted using Matlab™, and conclusions drawn about the impact of the inclusion on the crystal phase of PVF_2.

Investigation of Smart Materials

The subject of the internship is Smart Materials, those which exhibit a reversible property change as a response upon application and/or removal of a stimulus. Piezoelectric films constitute one type of Smart Material; the stimulus can be a change in temperature or pressure that changes the dimension of the piezoelectric film, and the response is flow of electrical current in an external circuit, as shown in **Figure 3**.

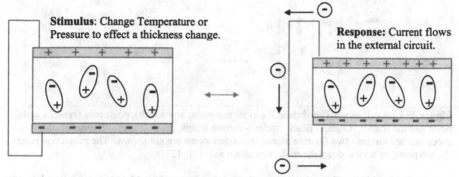

Figure 3. Example of an electrically active Smart Material. A piezoelectric film, when stressed by a change in thickness, causes an electrical current to flow.

The polymeric Smart Material chosen for the internship is poly(vinylidene fluoride), PVF_2 (also known as PVDF) which exists in several crystallographic phases, including non-polar alpha phase (tgtg') and the polar beta (ttt) and gamma (tttgtttg') phases, whose unit cells are shown in **Figure 4** [11,12]. Industrially, beta phase piezoelectric films are made by melt processing a thin film to obtain non-polar alpha phase, and then stretching in an electric field, an operation called "poling" to create the all-trans polar beta phase, whose electric dipole moments become oriented globally by the poling field. The objective of the internship work for 2004 and 2005 was to investigate the effect of clay addition on PVF_2/OMS nanocomposites that have been crystallized by heating from the glassy state (2004) or by cooling from the molten state (2005). In neat PVF_2 these treatments lead exclusively to growth of non-polar alpha phase [13-16].

Figure 5a,b shows example of the 2005 interns' results from X-ray and FTIR, respectively, for samples melt crystallized at 150 °C. In neat PVF_2 this treatment results in alpha phase, which phase diminishes as the OMS content increases. Room temperature composite WAXS scans are shown in **Figure 5a**. For homopolymer PVF_2 (bottom curve) alpha phase reflections with d-spacings $d_{100}(\alpha) = 0.49$ nm, $d_{020}(\alpha) = 0.48$ nm, $d_{110}(\alpha) = 0.44$ nm and $d_{021}(\alpha) = 0.33$ nm, are observed clearly at the 2θ angles (for λ = 0.154 nm) 17.66, 18.3, 19.9, and 26.7 degrees, respectively [17,18]. Two other reflections at higher angles appearing as shoulders on (021) peak, at 2θ angles 25.8 and 28.1 degrees correspond to $d_{120}(\alpha) = 0.34$ nm and $d_{111}(\alpha) = 0.32$ nm. Gamma crystal planes are known to have very similar WAXS reflections to the ones of beta and alpha phase crystals [19]. The peak on the lowest angle side with Miller indices (100) is the only peak among major alpha characteristic peaks that does not have any possibility of overlapping reflections with gamma phase. D-spacings of gamma reflections at $d_{020}(\gamma) = 0.480$, $d_{110}(\gamma) = 0.442$ and $d_{022}(\gamma) = 0.333$ nm overlap with (020), (110) and (021) alpha crystal planes, respectively. These overlapping reflections do not allow us to detect if there exists any small amount of gamma phase in neat PVF$_2$ from WAXS data. However, in FTIR data (shown in Figure 5b) we did not see any indication of gamma crystals for homopolymer PVF$_2$.

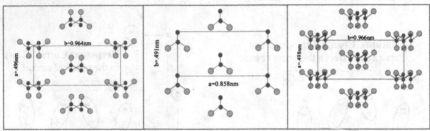

Figure 4. Unit cell of poly(vinylidene fluoride) non-polar alpha (left), polar beta (center), and polar gamma (right). Legend: Black circles – carbon atoms; green circles - fluorine atoms; green-slashed circles - two fluorine atoms; hydrogen atoms are not shown. The projection is into the a-b plane, in a view down the molecular chain axis [11, 12].

As clay is added to PVF$_2$, the (100) peak of alpha phase at 2θ = 17.66° starts to diminish, and vanishes for nanocomposites with 0.32% and 1.0% OMS. Similarly the peaks on the higher angle side show a decrease with introduction of OMS into the polymer leaving only a slight reflection at the highest PVF$_2$/OMS compositions. In samples containing the two highest amounts of OMS, the (100) peak of alpha is not observed which indicates that the peak at 2θ = 18.3° belongs to (020) gamma reflection for those compositions. The major peak at 2θ = 19.90° broadens and shifts to higher theta degrees, 2θ = 20.20° upon adding of OMS to PVF$_2$, as a result of new crystal phase introduction. This newly emerging peak could be attributed to beta phase characteristic reflection 200/110 with a d-spacing 0.427 nm, but there is also a gamma reflection $d021(\gamma) = 0.431$ nm in the same position [19,20].

Like beta crystals, gamma-PVF$_2$ crystals are polar, but are not usually obtainable except by high temperature melt crystallization for long times. Addition of OMS favors formation of polar forms of PVF$_2$ (beta after cold crystallization, and gamma with small amount of beta after melt crystallization) and may offer a new possibility for processing piezo-films without stretching. Poling in an electric field would still be required to orient the dipoles normal to the film plane.

Figure 5b shows FTIR spectra of melt-crystallized PVF$_2$/OMS nanocomposites from 400 – 1500 cm^{-1}. The absorption band assignment has been made by reference to the literature [21-24]. The bands listed across the bottom of Figure 5b are assigned as characteristic peaks of alpha phase, at 1383, 1215, 976, 796, 764, and 532 cm^{-1}, and are clearly observed in the bottom scan of neat PVF$_2$. The bands listed across the top of Figure 5b are the beta and gamma peaks. Gamma characteristic peaks occur at frequencies 1233, 883, 840, 811, 510, and 430 cm^{-1}. In these melt-crystallized nanocomposites, at 0.07wt% OMS and greater, gamma phase dominates [9].

Due to the existence of TTT in the chain conformation in both gamma and beta phases of PVF$_2$, most of the absorption bands for these two phases appear at the same, or very similar, frequencies in the FTIR spectrum: bands at 883, 840, and 510 cm^{-1} are also characteristic of beta phase absorption. The beta form of PVF$_2$ has other characteristic bands at 442 and 1275 cm^{-1} (marked at the top of Figure 5b), and 467 cm^{-1} (not marked). To identify gamma and beta phases uniquely, we must use bands that have no overlap between these phases. Accordingly, we use the 1233 and 811 cm^{-1} peaks to assign gamma phase and 467 and 1275 cm^{-1} to assign beta phase. In melt-crystallized samples, 811 and 1233 cm^{-1} gamma-bands increase at or above 0.05% OMS.

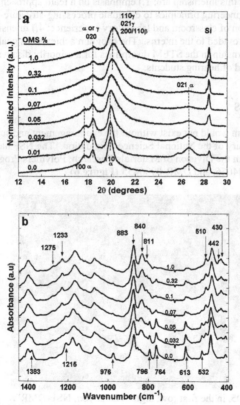

Figure 5 Characterization of PVF$_2$/OMS nanocomposites, melt crystallized at 150 °C for 1hr, with OMS compositions as indicated. a.) X-ray intensity vs. scattering angle, 2θ (for λ = 0.154nm). b.) FTIR absorbance vs. wavenumber. Curves are displaced vertically for clarity. In (b), prominent vibrational bands are noted [21-24]. Across the bottom are bands related to alpha phase PVF$_2$; across the top are bands related to beta or gamma phase PVF$_2$.

CONCLUSIONS

An internship program for deaf and hard of hearing students was organized around the topic of Polymer-based Nanocomposites, with application selected from the class of Smart Materials. Poly(vinylidene fluoride) is piezoelectric in its beta crystallographic phase, and the DHH interns from 2004-2008 explored a variety of inclusions (clay or carbon nanotubes) to characterize the crystal phase. The beta phase is favored by addition of Lucentite OMS for quenched samples [1-3]. Gamma phase is favored by addition of Lucentite OMS for melt crystallized samples [9]. Beta and gamma are both polar crystals, and the interns identified an attractive means of producing these phases, which could lead to new strategies for creating polar materials.

The unique aspects of this internship are: 1.) emphasis on a team approach; 2.) use of Materials Science and Engineering principles to elucidate processing-structure-property relationships; 3.) integration of classroom and laboratory experience; 4.) intensive daily mentoring and guidance provided to the interns. The program achieved its goal of providing a scientifically rigorous internship in the STEM disciplines for a minority disadvantaged population of deaf and hard of hearing students.

ACKNOWLEDGMENTS

The internship program would not exist without the encouragement and support of Andy Lovinger and Freddy Khoury of the National Science Foundation. The NSF has sponsored this program through 2011 from the Division of Materials Research, Polymers Program, under grants DMR-0406127, DMR-0704056, and DMR-0906455 (**Figure 6**).

Figure 6 Interns from 2005, in the first row, sign "Thank you, NSF-DMR".

This paper is dedicated to the memory of Prof. Robert Guertin, who participated in the internship program from 2004-2007, and passed away in June 2009.

REFERENCES

1. L. Priya, J. .P Jog, *J Polym Sci: Part B: Polym Phys,* **89**, 2036 (2003).
2. L. Priya, J. P. Jog, *J Polym Sci, Part B: Polym Phys,* **40**, 1682 (2002).
3. J. Buckley, P. Cebe, D. Cherdack, J. Crawford, B. S. Ince, M. Jenkins, J. Pan, M. Reveley, N. Washington, N. Wolchover, *Polymer,* **47**, 2411 (2006).
4. S. Ramasundaram, S. Yoon, K. J. Kim, C. Park, *J Polym Sci, Part B: Polym Phys* **46** 2173 (2008).
5. P. Cebe, D. Cherdack, R. Guertin, T. Haas, B. S. Ince, R. Valluzzi, *J Materials Education,* **28(1)**, 151-158 (2006).
6. P. Cebe, "A Team Approach: Using BEST Principles in an Internship Program for Deaf and Hard of Hearing Undergraduates." *Broadening Participation in Undergraduate Research.* Eds., M. Boyd and J. Wesemann (Council on Undergraduate Research: Washington DC 2008, 2009) Section III, 215-218.
7. http://www.chemistry.org/portal/a/c/s/1/acsdisplay.html?DOC=membership\index.html The American Chemical Society lists 158,422 members for 2005, with 25% female members. Self-identified ACS members with a disability (of any type, not just deaf or hard of hearing) numbered 339, or 0.21% of members.
8. http://www.nsf.gov/statistics/wmpd/listtables.htm National Science Foundation in Table B-10 gives the number of undergraduates in all Engineering Disciplines for 2002 at 421,178. Black undergraduates comprised 6.28%; females comprised 18.5%.
9. B. S. Ince-Gunduz, R. Alpern, D. Amare, J. Crawford, B. Dolan, S. Jones, R. Kobylarz, M. Reveley, P. Cebe, *Polymer,* in review (2010).
10. W. Huang, K. Edenzon, L. Fernandez, S. Razmpour, J. Woodburn, and P. Cebe. *J. Applied Polym. Sci.,* **115(6)**, 3238-3248 (2009).
11. A. J. Lovinger, "Poly(vinylidene fluoride)." *Developments in Crystalline Polymers–1.* Editor, D. C. Bassett (Applied Science Publishers, London, 1982).
12. A. J. Lovinger, *Science,* **220**, 1115 (1983)
13. S. Nakamura, T. Sasaki, J. Funamoto, K. Matsuzaki, *Makromol Chem,* **176**, 3471 (1975).
14. L. Mancarella, E. Martuscelli, *Polymer,* **18**, 1240 (1977).
15. H. Kawai, *Jpn J Appl Phys,* **8(7)**, 1975 (1969).
16. G. J. Welch, R. L. Miller, *J Polym Sci Polym Phys Ed,* **14**, 1683 (1976).
17. M. Bachmann, J. B. Lando, *Macromolecules,* **14**, 40 (1981).
18. J. B. Lando, H. G. Olf, A. Peterlin, *J Polym Sci A-1,* **4**, 941 (1966).
19. A. J. Lovinger, *Macromolecules,* **14**, 322 (1981).
20. R. Hasegawa, Y. Takahashi, Y. Chatani, H. Tadokoro, *Polym J,* **3**, 600 (1972).
21. M. Kobayashi, K. Tashiro, H. Tadokoro, *Macromolecules,* **8**, 158 (1975).
22. D. Yang, Y. Chen, *J Mater Sci Lett,* **6**, 599 (1987).
23. M. A. Bachmann, W. L. Gordon, *J Appl Phys,* **50**, 6106 (1979).
24. J. Kressler, R. Schafer, R. Thomann, *Appl Spectr,* **52**, 1269 (1998).

Mater. Res. Soc. Symp. Proc. Vol. 1233 © 2010 Materials Research Society   1233-PP10-03

Research Grade Instrumentation for Nanotechnology and MSE Education

[1,3] Christine Broadbridge, [1,3] Jacquelynn Garofano, [2,3] Eric Altman, [2] Yehia Khalil, [2,3] Victor Henrich, [2,3] Yaron Segal, [2] Myrtle-Rose Padmore, [2,3] Philip Michael and [2,3] Fred Walker

[1] Department of Physics, Southern Connecticut State University, New Haven, CT, USA
[2] School of Engineering and Applied Science, Yale University, New Haven, CT, USA
[3] Center for Research on Interface Structures and Phenomena (CRISP), Yale/SCSU

ABSTRACT

The Center for Research on Interface Structures and Phenomena (CRISP) is a National Science Foundation (NSF) Materials Research Science and Engineering Center (MRSEC). CRISP is a partnership between Yale University, Southern Connecticut State University (SCSU) and Brookhaven National Laboratory. A main focus of CRISP research is complex oxide interfaces that are prepared using epitaxial techniques, including molecular beam epitaxy (MBE). Complex oxides exhibit a wealth of electronic, magnetic and chemical behaviors, and the surfaces and interfaces of complex oxides can have properties that differ substantially from those of the corresponding bulk materials. CRISP employs this research program in a concerted way to educate students at all levels. CRISP has constructed a robust MBE apparatus specifically designed for safe and productive use by undergraduates. Students can grow their own samples and then characterize them with facilities at both Yale and SCSU, providing a complete research and educational experience. This paper will focus on the implementation of the CRISP Teaching MBE facility and its use in the study of the synthesis and properties of the crystalline oxide-silicon interface.

INTRODUCTION

The mission of CRISP Education and Outreach (E&O) is to use materials science as a vehicle for enhancing the scientific literacy of K-12 and undergraduate through post-graduate level students, educators and members of the community, and to improve the quality and diversity of science education for current and future scientists and science educators. The educational goals and resulting signature programs were designed to optimize integration of the research and educational strengths of our MRSEC through various E&O programs. CRISP E&O activities include (i) providing research experiences for local high-school teachers engaged in SCSU's Master of Science (M.S.) in Science Education program, (ii) development of specialized facilities for research, (iii) education and training in fabrication and characterization instrumentation, as well as (iv) team-based research opportunities for undergraduates, teachers and high school students. CRISP encourages students to pursue science in higher education, especially individuals from under-represented populations, while offering both informal and formal outreach to stimulate science literacy for future policy makers and the general community.

CRISP has constructed a robust molecular beam epitaxy (MBE) apparatus specifically designed for safe and productive use by undergraduates, graduates, post-docs and faculty. Students have the opportunity to be engaged in a complete research experience where they grow their own samples with MBE and then follow-up with characterization at facilities at Yale and the CRISP NanoCharacterization facility at SCSU. Complex oxide MBE is a primary research tool for CRISP, whose facilities were developed by pioneers in the oxide MBE and oxide characterization fields.

The goal of the CRISP Teaching MBE project was to build a state-of-the-art research MBE system allowing undergraduate students to learn first-hand about cutting-edge research by engaging directly with the determination of key deposition, measurement and characterization tools essential to MBE. Moreover, we have extended the use of the MBE to several of our programs in order to introduce nanofabrication by MBE as an integral part of materials science education. Such programs include Research Experiences for Undergraduates (REU), Research Experiences for Teachers (RET), and research-intensive M.S. courses and professional development (PD) workshops for educators. This paper will focus on the implementation of the CRISP Teaching MBE facility and the use of this instrumentation for undergraduate courses and research as well as a newly implemented PD workshop entitled *Nanofabrication: MBE*.

EXPERIMENTAL METHODS

MBE is a versatile fabrication technique for depositing thin films with desired composition and structure. The goal of the project was to use a state-of-the-art research MBE system to learn about and characterize the key deposition, measurement and characterization tools essential to epitaxial growth. Students investigated the growth of a template layer for crystalline $SrTiO_3$ on Si, an enabling system for faster computers and new generations of electronic devices. Specifically, students mapped out the boundaries of the two-dimensional surface phase diagram for Sr on Si by monitoring surface diffraction patterns as a function of Sr surface concentration and temperature [1].

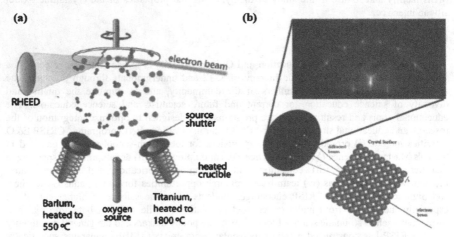

FIG 1. Schematic illustrating **(a)** MBE growth of $BaTiO_3$ [1] and **(b)** *in situ* characterization of film growth by reflection high energy electron diffraction (RHEED) [2].

Although the term *molecular beam epitaxy* may suggest a complex process, the general concept is straightforward, as shown in Fig. 1. In MBE, materials that do not have significant vapor pressures near room temperature are heated to high enough temperatures that they sublime or evaporate. If a cooler substrate is placed in the path of the vapor emanating from the heated material, then that material will condense on the substrate and a film will begin to grow. The experiment is carried out under vacuum such that the mean free path of the atoms and molecules in the gas phase, l, is much greater than the distance between the heated source material and the sample.

The mean free path is given by:

$$l = \frac{kT}{\sqrt{2}\sigma P} \qquad (1)$$

where k is Boltzman's constant, T the temperature (K), σ the cross-sectional area of the atom or molecule (typically 0.07 nm^2), and P the pressure (Torr). At 10^{-6} Torr the mean-free-path is 300 meters (m), and so in any realistic vacuum system, the atoms or molecules emanating from the source will not have an opportunity to collide with other gas molecules before hitting the substrate. The plume emanating from the source is thus considered a "molecular beam." This approach has several advantages, including (i) the potentially high purity of the vapor emanating from the source, (ii) high thermal efficiencies due to the lack of convective heat losses, and (iii) the need to heat only a small source to high temperatures rather than an entire enclosure. Although pressures in the 10^{-4} Torr range would be sufficient to ensure that the mean free path is greater than the distance between the source and substrate (typically ~ 0.3 m), MBE systems operate under ultra-high vacuum (UHV) conditions (< 10^{-9} Torr) to minimize contamination by adsorption or reaction with undesirable gas molecules [2].

The evolution of the surface structure of the epitaxial film is monitored *in situ* with reflection high energy electron diffraction (RHEED). In RHEED, electrons with energies on the order of 10 keV are scattered off the sample at grazing incidence and imaged on a phosphor screen. At these high energies the electron wavelength is less than the spacing between atoms in the solid, and so diffraction patterns due to constructive and destructive interference of electrons scattering off evenly spaced rows of atoms on the surface will appear. Using the grazing incidence has two of advantages. First, the electrons will not penetrate deeply into the solid, so the diffraction pattern will reflect the crystalline order of only the outermost one or two layers, allowing changes in the structure of the growing film to be monitored in real time. Second, the electron source, or gun, and phosphor screen can be placed at opposite sides of the vacuum chamber, out of the path of the molecular beam.

The concept of epitaxial growth is important to this experiment. The word epitaxy comes from the Greek *epi*, meaning "onto", and *taxi*, meaning "order". Thus, epitaxy describes the deposition of an ordered film onto a substrate. More specifically, it is generally taken to mean that the film is crystalline, with a well-defined relationship between the crystallographic axes of the substrate and the film. When atoms are initially deposited onto a substrate at high temperatures, the coverage (or density) is low and the film atoms fall onto random sites on the surface; since the temperature is high, the atoms may diffuse between sites, similar to the behavior of a gas. Because the film atoms are restricted to well-defined sites, this initial phase is referred to as a "lattice gas," as shown schematically in Fig. 2(a). As the coverage increases, we may expect the film atoms to "condense" into an ordered structure (Fig. 2(b)). If the coverage is too low for this condensed phase to cover the entire surface, it will be in equilibrium with a lattice gas. Increasing the coverage beyond the maximum defined by the structure of the condensed phase can cause either a second layer of film atoms to form, or a new, higher density condensed phase to form, which would be in equilibrium with the lower density phase.

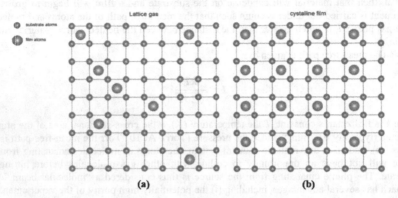

FIG. 2 Schematic representation of the formation of **(a)** the lattice gas at low coverage where the red atoms occupy random sites on the surface, and **(b)** an ordered, crystalline film as coverage increases and condensation causes a higher density solid phase to form [2].

IMPLEMENTATION

The CRISP Teaching MBE has been used for course modules, research experiences and, most recently, professional development. For this paper, we will focus on the use of MBE during an undergraduate course at Yale entitled *Chemical Engineering Laboratory* with the use of the lab module: *From Surface to Interface: The Synthesis and Properties of a Crystalline Oxide-Silicon Interface* [2]. Students were given the following tasks: (i) growing $SrTiO_3$ on Si, (ii) mapping the boundaries of the two-dimensional (2D) surface phase diagram for Sr on Si, and (iii) determining the heat of condensation of the ordered phases. The heat of condensation for the formation of the solid 2D phase from the lattice gas can be obtained from the Clausius-Clapeyron equation by measuring the coverage θ at which the 2D phase forms as a function of temperature. That is, at equilibrium, the free energy change in going from the lattice gas to the 2D solid must be zero, and therefore

$$T\Delta S = \Delta H \tag{2}$$

where T is temperature, S is entropy and H is free energy. The entropy of a lattice gas is related to all the possible ways to arrange m atoms onto n sites. It can be shown that this reduces to:

$$\Delta S = k \ln\left(\frac{\theta}{1-\theta}\right) \tag{3}$$

where k is again Boltzman's constant. Therefore, substituting **(2)** into **(3)** yields

$$\Delta H = kT \ln\left(\frac{\theta}{1-\theta}\right) \qquad (4)$$

and the heat of condensation can be measured by determining the coverage at which the 2D solid forms at different temperatures.

Students were asked to determine the phase diagram for Sr deposited on Si. The procedure was as follows: a Si wafer was inserted into the chamber and heated to desorb any SiO_2 from the surface. RHEED diffraction patterns (DPs) were obtained with the sample aligned at various angles with respect to the electron beam. Students observed how the DPs changed with respect to angle. Once picking an appropriate angle and sample temperature, DPs were monitored as a function of deposition time, and thus coverage, assuming all of the Sr that hits the surface sticks. Finally, the sample was heated to a higher temperature to desorb the Sr and the experiment was repeated at several other temperatures.

EXPERIMENTAL RESULTS

Experiments for Sr deposition onto Si(100) have revealed the presence of at least four phases, a lattice gas phase and three ordered phases: (3×2), (2×1), and (3×1). Based on the sequence in which these phases were observed as the coverage was increased, the approximate phase diagram in Fig. 3(a) was created. Students were able to determine the transition from disordered lattice gas, superimposed on the substrate (2×1) structure, to ordered (3×2) condensate on the film surface, as shown in Fig. 3(b), which plots the submonolayer intensity oscillation of a diffraction peak associated with the (3×2) phase as a function of time (s). At 625 °C and a cell temperature of 350°C, the phase transition was complete at 1/6 monolayer (ML) coverage.

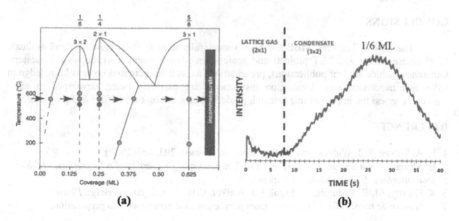

(a)   (b)

FIG. 3 (a) Approximate 2D phase diagram for Sr on Si(100), where the points represent experimental data [2]; (b) plot of submonolayer intensity oscillation versus time reveals the phase transition from lattice gas to ordered condensate begins at 8 sec and is complete at 1/6 ML at 27 sec, indicating that the maximum density of the lattice gas is 0.05 ML at 625°C [3].

21

PROGRAM RESULTS

Depending on the format of the MBE module, student participation varied from direct hands-on data collection and analysis (e.g. REU and RET) to combined demonstration/hands-on for professional development workshops. Course modules provided as much hands-on data collection as feasible given constraints including class time, class size and course objectives. Undergraduate courses with substantial MBE modules include: *The Science of Nanostructures* physics course at SCSU and *Chemical Engineering Laboratory* course at Yale. In both cases, students collect and interpret data directly.

The MBE module has been used extensively by undergraduates in their research and for graduate students in either Science Education (SCSU) or Applied Physics (Yale or SCSU). For instance, as a result of an academic year undergraduate research project, a comprehensive study was undertaken culminating in a comparison between the experimental data and theoretical (density functional theory) calculations for the Sr to Si interface [3]. With respect to M.S. courses, a research-intensive course for science teachers was developed and has been offered twice; a second course (*Nano-synthesis and Fabrication*) is currently under development for the newly established M.S. in Applied Physics program at SCSU. This course will be offered during the summer of 2010 and will use the CRISP Teaching MBE extensively.

Most recently, the MBE module was used for professional development during a six-hour PD workshop entitled *Nanofabrication: MBE* provided to eight educators [grades 9-14] at Yale. The goals of this workshop included education and training in nanofabrication and to exploit the interdisciplinary nature of MBE, materials science and nanotechnology. Results of the workshop assessment survey indicated that, while educators enhanced their knowledge and understanding in all areas, their primary benefit was an increased understanding of how to align nanotechnology and materials science with their curriculum.

CONCLUSIONS

The CRISP Teaching MBE has been successfully used for undergraduate and graduate level courses, REU and RET projects and professional development workshops for educators. Outcomes include student publications, presentations as well as increased content knowledge in MSE and nanotechnology. Evaluation instruments are currently being developed to more rigorously assess the impact of this research grade instrumentation on CRISP E&O programs.

REFERENCES

1. R. A. Mckee, F. J. Walker, and M. F. Chisholm, *Science*, **293**, 468(2001).
2. E. Altman, *From Surface to Interface: The Synthesis and Properties of a Crystalline Oxide-Silicon Interface,* MBE module description.
3. K. Garrity, M.-R. Padmore, Y. Segal, F.J. Walker, C.H. Ahn, S. Ismail-Beigi, *Phase Transition of Sr on Si (001): First-principles prediction and experiment;* in preparation.

ACKNOWLEDGEMENTS

This research and the educational programs described are supported by NSF Grant MRSEC DMR05-20495.

Mater. Res. Soc. Symp. Proc. Vol. 1233 © 2010 Materials Research Society   1233-PP07-01

Innovative Evaluation of Two Materials Science Education Enrichment Programs

Daniel J. Steinberg, Shannon L. Greco and Kimberly Carroll
PCCM, Princeton University, 316 Bowen Hall, 70 Prospect Ave., Princeton, NJ, 08544, U.S.A.

ABSTRACT

The Princeton Center for Complex Materials (PCCM) is an NSF-funded Materials Science and Research Center (MRSEC) at Princeton. PCCM currently has four Interdisciplinary Research Groups (IRGs) and several seed projects. PCCM runs a variety of education outreach programs that include: Research Experience for Undergraduates, Research Experience for Teachers, Materials Camp for Teachers, Middle School Science and Engineering Expo (SEE) for 1200 students, and Princeton University Materials Academy (PUMA), for inner city high school students. In this paper we focus on new evaluation efforts for the PUMA and the Science and Engineering Expo. We will discuss first PUMA the SEE and elaborate on the new evaluation efforts for each program.

Created in 2002 by PCCM, PUMA has an inquiry based materials science curriculum designed to work at the high school level. PUMA's activities are paired with an inquiry based evaluation of scientific ability and attitude change. An evaluation of high school students' ability to formulate scientific questions as a result of their participation in this summer program based was developed based on similar studies of college students questioning ability in inquiry learning environments. Created in 2004 by PCCM and partners in Molecular Biology, SEE is run once per year in the spring. It is a day dedicated to capturing the imaginations of young students through science demonstrations and direct interaction with materials scientists and engineers. 1000 middle school students from local schools come to Princeton University to interact with Princeton scientists and engineers and explore science with the help of demonstrations and hands-on activities. Throughout the day, they explore a wide range research from Princeton that is at the cutting edge of science and engineering to generate excitement about science and engineering. In addition to studying over 5000's student written essays we have constructed a pre and post test for student attitudes administered to over 500 students in 2009 to determine the impact of the SEE on students' attitudes about materials science and STEM fields. This large scale attitude assessment and student written statements help to establish the impact of this one day program.

PRINCETON UNIVERSITY MATERIALS ACADEMY (PUMA)

Program Description

Princeton University Materials Academy (PUMA) is summer enrichment program designed to introduce inner city Trenton Central High School students to material science and engineering. Each year, the students learn about materials science related topics such as energy, climate change, BioMEMS, and water filtration. The daily schedule is filled with lectures, lab activities, and challenges to help them better understand the science behind the research they are performing. Princeton faculty, staff, and students contribute their time to conduct lectures, lab activities, or develop new curriculum based on their work. PUMA applies a project-based learning approach in science and inquiry learning. Students in PUMA are primarily from underrepresented minority groups, from low income families, and a majority of the students have been female each year.

Students learn about their projects from a material science standpoint. For example, rather than learn about solar ovens only in terms of physics and energy conservation, the whole project is looked at from all the different disciplines that material science combines. Students are taught that material science is multi-disciplinary and then are able to see that this is true through the projects, lectures and labs in which they participate.

According to the students, some of them stated that they had never taken a science class with labs prior to entering PUMA. They had no experience with having to ask questions in order to further their research, nor had they have experienced a situation where there was more than a single answer to a question. PUMA gives these students the opportunity to experience labs, as well as talk to experts in the field. Through these labs and interactions with experts, the PUMA students are exposed to a great deal of information, sometimes not in a linear way, but the education director works with the master teacher to help the students categorize new knowledge and reflect on what they have learned.

At the end of the program, students are able to ask scientific questions of the PUMA subject matter and of the world around them. Most of all, PUMA offers a program that is comprehensive, understandable and appropriate for the level of the students and helps the students to think scientifically, which is evident through their increasing level of scientific questions. [1]

As the program is funded by the National Science Foundation, an emphasis on some of NSF's broader impact goals is prevalent. The broader impact that is most addressed in the program is the creation of scientifically literate community, both within and outside the science community. PUMA does this by introducing inquiry to students so that in the future, regardless of career choice, will question everything they see and read.

Background

The research into fostering students' ability to ask meaningful scientific questions via inquiry based learning was not new when it was taken up by PCCM, but actually testing it out on high school students with little scientific background was a new idea. There has been research on students' questions but it has been focused on university students, who typically have a different background than an inner city high school student (*Harper et al 2003*). Despite this difference, it was possible to apply many of the major aspects to the PCCM research project.

This research was the basis for determining if questions could be used to measure PUMA's impact on student questioning ability. If students can think critically about the material, question the material and truly think like scientists, then PUMA is giving them some lab experience that will impact them greatly, especially the students who choose to pursue sciences. By measuring questioning ability, PUMA is being evaluated as a program on its ability to teach this skill. Through instruction, reflection and remediation, PUMA aims to teach students what a scientific question is and to develop students' skill in asking scientific questions.

The project served more than just an evaluation tool, however. The questions force students to think about what they are learning, which is a skill called "metacognition". If students are thinking critically about what they are learning, their ability to ask questions and what knowledge they might gain from asking the questions, the hope is that they will internalize their knowledge more effectively and will have a deeper understanding of the material. This, too, is an important part of thinking like a scientist that often gets lost in a standard classroom.

Procedure

A question rubric was made based on Harper and Etkina's university-level rubric.[2-5] Questions could be scored from 0 (minimal level) to 3 (high level). Scientific questions were defined as a question that shows an understanding of material, can be answered with an experiment and contains enough information that there is no confusion about the direction an experiment might go towards. The rubric scored students on a sliding scale of these major contributing factors.

In the first session, students were told nothing of the experiment but were asked to respond to three question prompts: If you were teaching this class, what scientific questions would you like to students to ask to make sure they understand what we have been learning? What questions do you still have about what we have been learning? What questions do you still have about material science? The questions were then collected and scored by evaluators.

Following this session, students were given copies of the rubric and practiced evaluating questions that they came up with on the scale. Most students agreed on their evaluations. Students were then taught what a scientific question is and why it is so important.

Subsequent sessions had the same question prompts, but students were reminded to think about their question before recording it. These questions were also collected and scored by a team of evaluators. The scoring rubric had high inter-rater reliability among the three raters performing this research. Hands-on activities were interspersed in the program where students could find answers to their questions and interact with real scientists who told students of their own scientific questions, how they had devised them and what they were doing to answer their questions. Most of these scientists helped students with lab projects and experiments, offering help with what questions might be helpful to ask and what information might be useful to know.

Results

On a 0-3 point scale below are the results for 2007, 2008, 2009. Results were found by scoring each student question and then finding the class average.

Table I, Class averages by session:

	2007	2008	2009
Session 1	0.39	0.56	0.33
Session 2	0.87	0.69	0.57
Session 3	1.14	1.06	0.8
Session 4	0.96	0.42	0.96

This shows that each year, improvements in questioning ability were made. Question improvements tended to be made after the remediation (as evidenced by the session 1 to session 2 changes). In years 2007 and 2008, dips in level were found between sessions 3 and 4. This could be due to a lack of remediation and availability of scientists not allowing for visits in these times. It is safe to say that according to the above results, levels of questions improved with the combined effort of scientist interaction, inquiry based activities and question remediation.

Improvements

As with any innovative research, changes can be made to improve the quality and reliability of the results. In all 3 years of the study (2007-2009), students were required to ask questions, but it was found that results were more consistent among individual student progress when only students who were "getting it" and trying hard to apply good scientific questions as opposed to a few who were merely writing something down to get an "assignment" done In the future, students should be encouraged to write out questions, but participation will be voluntary.

Remediation

Because students have never been exposed to scientific questions, they do not know how to ask one. This means that they cannot reasonably be expected to ask high level questions unless they have the concept of questioning explained to them. It becomes necessary to teach the students about questions, simply because otherwise, the study is meaningless, as the term "scientific question" is not a part of their understood vocabulary. The students are taught the criteria for a good question and given an in class, informal assessment conducted through competitive games.

For each year the study was conducted, the idea of questions as a form of assessment was not well received by students initially. Students are used to being asked factual questions to which they must provide an answer to get a grade. They have learned that there is always a right answer. With what we were doing, the students were told at the start that they would not be graded and that nothing was "wrong".[6] This is a surprise to most students and can easily be viewed as an inaccurate representation of the real world by students because of its inconsistency with their own experience. This is, however, not a surprising roadblock, and it is this precise attitude PUMA is meant to change – to show them that text book science is not necessarily an accurate representation of the real world and that scientists work with unanswered questions regularly.

One issue that must be addressed is how the numbers dipped for the last session in years 2007 and 2008, though no such dip was observed in 2009. These dips in the 1st two years can be explained by the reasons set forth. In between the third and fourth sessions, both years, there was a significant dip in new content since students were wrapping up their projects and working on their final presentations. There was also some lack of teacher buy-in. In 2009 the program was expanded to 3 weeks and new content continued into the 4th questioning session. Also in 2009 the teacher buy in to the program greatly improved. Both of these factors seem to have improved the resulting student questioning ability as measured by the rubric.

PUMA Conclusions

PUMA is a unique program that engages and enables at-risk students in science who otherwise may not have a chance to succeed. Although there are well respected and very good research programs for high school students, the students in PUMA would not have the skills to participate without the help of PUMA. PUMA teaches students how to think in the way scientists do – questioning what they are told and being critical of what they see. This skill is essential for going into sciences in graduate school, which some PUMA students do and it is important for any citizen to have, with science becoming a greater, more visible influence in their daily lives. The improvements shown in students' questions indicate that PUMA is able to enhance metacognition and critical thinking in science, which fulfills the goals of PUMA, PCCM and NSF.

From three years of data (Table I), it has been shown that high school student questioning ability can be measured. The resulting improvements are at least as striking as those shown in studies of students questioning abilities in introductory college courses. PUMA is an inquiry based program, where scientific questioning is valued, explicitly taught, and practiced, resulting in improved student ability to ask scientific questions. Not all students were willing participants, and this negatively impacted the results of the study. Some students wrote nonsense questions that were merely a way to get credit for having done the work, but not actually focus on true megacognition. Since the major goal of this project was to see if measuring questions was a reliable and valid form of evaluation, it was not imperative to ensure that all scored questions were from serious students, since class averages did show improvement. However, in the future, having established that PUMA and question remediation do positively impact students, participation in the study can be voluntary in an effort to measure the true gains by focusing on students who put sufficient effort into every question that they ask. It is hoped we can prepare these students for more sophisticated research opportunities and that they will explore Research experience for Undergraduates programs when they are enrolled in college.

In 2007, we presented a paper – a cookbook of how to develop a MRSEC based Science Expo that could reach over 1000 middle school students. This is a discussion of our in-house evaluation efforts of our own program, the Science & Engineering Expo. We did work with an external evaluator, Magnolia Consulting, to confirm the validity of our pre- and post-surveys, which were developed based on the research of museum and other science education evaluators. We strive to meet broader impact guidelines set by our funders (National Science Foundation) by designing programs that maximize the science content expertise and enthusiasm of MRSEC faculty and science. With our evaluation efforts, we aim to improve the programs and to show a positive impact to prove our programs work to both NSF and to ourselves. This positive impact can be shown through improved attitudes towards science and improved understanding of science and engineering.

A new evaluation plan

In 2008, Steinberg hosted the MRSEC Education Directors' Network meeting at Princeton University at which the group focused on Logic Models for program development and evaluation. The group worked with Magnolia Consulting to improve the process of making logic models. Princeton Center for Complex Materials (PCCM, Princeton University's MRSEC) regularly employs logic models to explicitly state desired outcomes, what we are doing to work towards those outcomes, and how we know we have achieved those outcomes. The process of the logic model led us to develop pre- and post-surveys to measure impact on students' attitudes towards science and engineering, scientists and engineers, as well as a few other social issues related to science.

We believe we have the ability to measure at least four of our anticipated outcomes. We believe we increase student content knowledge and skills in MSE and STEM in general. We believe we have been measuring that every year of the SEE since 2004. From the conception of the program, wanted to prevent loss of positive attitudes towards science and later became focused on improve attitudes. We focused on middle school because middle school is a critical age at which many students lose interest in science.

In 2009, we added the pre- and post-survey, a summative component to measure the increased excitement and interest in MSE/STEM, increased exposure and awareness of MSE/STEM fields and careers, and an increased understanding of their own potential as scientists and engineers. From the first year, our major evaluation techniques have been a formative observation by organizers, scientists and volunteers. We observe that most students and many teachers are actively engaged in the event spending time at the tables, communicating with the volunteers, asking questions, manipulating hands-on activities, and observing science phenomena. While facial expressions are difficult to quantify[7], organizers and volunteers observe smiles and "expressions of wonder." These observations have confirmed that most of our table activities are highly successful in engaging students in science and engineering. Since we invite the schools, we have direct communication with the teachers and teachers have honored our request to have each and every student write a testimonial of their experience, sometimes the following day, sometimes weeks later. Over 90% of the students send us these testimonials and we have read all of them. In addition to the full attendance each year and the fact that we have more requests than we can handle, every school has wanted to return each year.

Testimonials are a strong part of our evaluation measuring content learning and attitude improvement. In reading testimonials, which are overwhelmingly positive, students report that this event an extremely positive experience for them. Students often cite specific scientists, experiments, and concepts. In their testimonials, they name and activity or scientist as their "favorite" or "the best" and often express desire to learn more. These testimonials contain valuable information that we plan to research further.

"If information is not worth recognizing, the memory of that data may be discarded."[8] We believe the students remembered what made an impression on them, which is evidence of learning. A male student from Montgomery Middle School (SEE 2009) cites two things he learned at the event in his statement, "Global warming has spiked the most in the past 40 years. Algae makes a lot more ethanol than corn does." A female student from Montgomery (SEE 2009) states, "The Carbon dioxide emission in the past 100 years has been increasing a great deal." Despite an effort to reach middle school content benchmarks, some misconceptions were evident in their testimonials, yet excitement and curiosity about the content are clearly expressed. Student statements overwhelmingly indicated, often explicitly, an enjoyment of meeting real scientists and participating in table activities. Many students cite what they learned and are accurate enough to be able to ask impressive questions. For example, another male student from Montgomery Middle School (SEE 2009) writes, "A superconductor can carry an electrical current without any resistance meaning that it keeps the current forever. Though it is ceramic and not a metal, it can carry an electrical current very well. Superconductivity only works at very low temperatures, so liquid nitrogen is used to cool the ceramic." He even asks "Why do superconductors only work at low temperatures? How is the structure of the superconductor changed when it reaches this low temperature?" This indicates not only and interest, but also understanding of some basic Materials Science.

From the student statements, we can see evidence of a renewed positive attitude and interest in science and engineering. We believed we were succeeding in improving these attitudes but to capture that, we needed to design a new evaluation plan. Our first attempt is the pre- and post-survey started in 2009. It is harder to separate content learning from their formal experiences, and content knowledge acquisition may be beyond our measurable outcomes. The statements and the logic model led us to the relevant literature in science and engineering education and the idea to pursue and develop the attitude survey.

Because our treatment is analogous to school field trip to a science museum, much of our literature search was in the field of informal science education. From Learning Science in Informal Environments, "Programs for science learning take place in schools and community-based and science-rich organizations and include sustained, self-organized activities of science enthusiasts. There is mounting evidence that structured, nonschool science programs can feed or stimulate the science-specific interests of adults and children, may positively influence academic achievement for students, and may expand participants' sense of future science career options."[7]

Participants Analysis

While the program's target audience is middle school students from NJ, the students are from varied backgrounds. The students come to us from school districts that are diverse in socio-economic status, cultural background, family education history and students with a variety of disabilities. This is by design and helps to create a fully inclusive atmosphere. Conversely, we have the same sort of diversity within the PCCM community which generates meaningful interaction with people from different cultures, backgrounds, and unique qualities.

Our female participation matches population percentage and our minority participation exceeds the percentage represented in the general population. Each year, of the 15 schools we invite, 5 are from Trenton (a school district that serves a 98% minority population and only 33% of 9th graders will graduate high school) and one is a school that specializes in students with disabilities. We ensure the event it fully accessible for physically disabled students. Since the program is strictly for students in grades 6 – 8 content is grade appropriate.

Connection of Activities to Outcomes

At the Dillon Gym venue, every student has free choice of which table activity and faculty member to interact with. One of our desired outcomes is to energize *all* students to someday consider STEM fields by showing them the true diversity within STEM. We have a great deal of impact on students' attitudes about STEM and provide a rich experience. We hope to increase that impact as we compete for more resources.

The duration of the program is designed for this age level. Students are given 30 minutes to explore the free-choice learning venue. The average time a visitor might spend at an activity in most free-choice learning environments is less than a minute. Free-choice learning environments allow the student to determine how much time is spent and is largely dictated by his or her attention span and engagement in the given activity. Many of our activities hold the attention of the students for 5 minutes or more.

Measurable Outcomes

As shown by our logic model, our short-term outcomes reflect reasonable, progressive steps towards are larger goals of what NSF calls "broader impacts." Research shows that episodic exposure to and engagement in STEM fields has a cumulative positive impact on students' interest and performance in STEM fields. We impact long-term outcomes in providing a key experience in what we hope is a rich series of positive experiences from many sources contributing to the scientific literacy and increased participation in the sciences. The outcomes we chose to measure are specific and can be measured by an attitude survey and essay responses. These instruments show that within the period of one day interest, excitement, awareness, personal potential to succeed in and content knowledge increased for the fields of materials science and engineering.

Opportunity to do more

We not only control the venue and content, we control the student participants. We have a good relationship with teachers and school administration, which gives us the ability to administer a pre- and post-test. For many programs, follow up is difficult but with these relationships already established, it is possible to hand teachers surveys and have almost every single survey returned. Post-surveys were given directly to the teachers in a pre-addressed, stamped envelope for completion and mailing after the event.

Enough surveys were printed for each student. In one package per school, the pre- surveys were mailed or hand delivered to one contact person at each school, with an instruction sheet asking teachers to have the students complete the pre-survey before the event and to bring the completed surveys to the event itself. An information booth was set up at the front of the Dillon Gym to collect completed pre-surveys.

Survey

We used the Test of Science-Related Attitudes (TOSRA)[9] as a starting point to develop our pre- and post-surveys on attitudes about topics covered in the Science & Engineering Expo. Using a similar classification of questions, we categorized our questions as follows: social implications of science; normality of scientists; attitude to scientific inquiry; adoption of scientific attitudes; enjoyment of science lessons; leisure interest in science; career interest in science; interest in materials science & engineering; awareness of climate change; and perception of gender equality in science.

The questions included on both the pre- and post-surveys were identical. Students were asked to rank each on a Likert scale ranging from Strongly Disagree (SD) to Strongly Agree (SA). In the analysis of the data, values were assigned to these ratings as follows: SD=1; D=2; N=3; A=4; SA=5. For negatively worded questions, the scale was reversed: SD=5; D=4; N=3; A=2; SA=1.

Our observations indicate that we did have a positive impact on excitement, interest, and motivation in science and engineering. We showed an increase in the degree to which students see themselves as scientists or potential scientists. We have designed the program to achieve these positive affective outcomes and designed our instruments to measure them.

The National Research Council's (NRC) Committee on Learning Science in Informal Environments outlined six strands of science learning, "statements about what learners do when they learn science, reflecting the practical as well as the more abstract, conceptual, and reflective aspects of science learning." The strands most relevant to this work are "Strand 1: Learners in informal environments experience excitement, interest, and motivation to learn about phenomena in the natural and physical world," and "Strand 6: Learners in informal environments think about themselves as science learners and develop an identity as someone who knows about, uses, and sometimes contributes to science." These two strands were added to an original 4 outlined in a previous NRC report to illustrate how informal science can compliment formal science education.[7] The Science & Engineering Expo is analogous to informal science learning environment therefore these statements can be made of the Expo as well.

Results

About 60% of the students whose teachers were given the survey completed both the pre- and the post-survey. Some teachers were given the survey and did not give it to their students. Some forgot to have their students complete the pre-survey before the event, yet completed the post-survey. This was the case for entire classes or schools and could therefore be eliminated from the post-survey data with confidence. For the classes that responded for both the pre- and the post-survey, we included their responses in our final data set. The pre-survey had 329 respondents and the post-survey had 320 respondents. The 9 students who did not complete the pre-survey could not be removed from the data set with confidence because the surveys were anonymous and no identifier was assigned to the students.

For the entire responding sample of the students participating in Science & Engineering Expo, we showed a significant positive impact four of our survey questions. These questions had a p-value of less than .05, allowing us to reject a null hypothesis. These questions were 3, 9, 13, and 16.

For question 3, "Doing experiments is not as good as finding out information from teachers," the results showed a shift towards "Strongly Disagree." The mean shifted from 4.21 (standard deviation of 0.91) to 4.02 (SD 1.11), meaning the students felt doing experiments was as good, if not better, than finding out information from teachers, $t(645) = 2.41$, $p = .02$. While we did not wish for students to dislike learning from their teachers, we did hope that students would prefer to find things out for themselves through hands-on, inquiry-based experiments. Some of our table activities, such as designing paper airplanes and roller coasters with returning PUMA students (see PUMA section), allowed students to test things for themselves.

Question 9 stated, "I am curious about polymers." Most students were not sure of their agreement with this statement before the treatment (mean of 3.04, SD 1.00), and after the treatment, more students agreed that they were curious about polymers (shifting the mean to 3.23, SD1.07), $t(634.95) = -2.32$, $p = .02$. This was an especially significant result as it was very specific to the topic of one particular table activity. To show that one table had such an impact on our results from the sample as a whole is very encouraging. One male student from Montgomery middle school states, "Polymers are substances made up of small particles bonded together to make larger particles. When water is added to a polymer, it links the polymer in larger particles," supporting our positive results from this survey question. This table is one in which we put the most effort in developing the content, activities, and interaction between the engineers and the students. Several years ago, PCCM education outreach worked with Liberty Science Center to develop the script and activities, and PCCM has made many improvements since then to perfect the script and make it more about the engineer's expertise. The polymer table had a PCCM faculty member and more grad students and postdoc volunteers than any of our other table activities. It is also suite of activities for which the education outreach director has worked extensively with this faculty member to perfect and he has practiced delivering the material in several of our education outreach programs.

For question 13, "Materials science may help solve our energy problems," the mean shifted from 3.81 (0.18) to 3.96 (0.89), $t(641) = -2.26$, $p = .02$. Students already agreed with this statement but shifted towards "strongly agree." A theme of the 2009 Science & Engineering Expo was Energy and a theme of every Science & Engineering Expo is materials science. We feel our efforts in communicating this are reflected in the results of our survey. Tables such as "Biofuels from Algae" and "Photovoltaics" highlighting our faculty's research helped make an impression on the middle school students.

On the normality of scientists, question 16 stated, "Scientists are less friendly than other people." Our hope was to shift students' attitudes towards scientists themselves as "normal" people just like the students themselves. If the scientists and engineers are seen as less different, then the hope is that the students will see their own potential to be a scientist. This statement showed a shift towards "Strongly Disagree" from a mean of 3.82 (SD 0.94) to 4.00 (SD 1.03), $t(645) = -2.32$, $p = .02$. Question 31, "I believe I have the ability to become a scientist," did show some positive change from the pre- to the post-survey, but with less statistically significant results (shift of mean 3.37, SD 1.18, to mean 3.51, SD 1.18, $t(640) = -1.55$, $p = .12$). Student statements further support our results showing positive impact in career aspirations in science, positive attitude toward scientific inquiry; adoption of scientific attitudes, interest in materials science & engineering and awareness of climate change. A male student from Timberlane Middle School (SEE 2009) writes, "The demonstrations were not only cool but inspiring. I intend to become a scientist with the goal of creating a perpetual motion machine to solve the energy crisis using magnets. My blue prints are below," and a drawing was included (proprietary information). We hope to show a greater improvement in this area in the future.

Cumulative Effects

As noted by the NRC, "science learning, and informal science learning more specifically, is a cumulative process. The impact of informal learning is not only the result of what happens at the time of the experience, but also the product of events happening before and after an experience. And interest in and knowledge of science is supported by experiences in informal environments and in schools."[7] We believe that our short term outcomes, when part of a collection of positive science learning experiences, lead to the intermediate term and then long term outcomes, ultimately contributing to the Broader Impacts as set by the National Science Foundation.

REFERENCES

1   Project 2061 (American Association for the Advancement of Science): *Benchmarks for science literacy*, New York, Oxford University Press, 1993.

2   Harper KA, Etkina E and Lin Y: Encouraging and Analyzing Student Questions in a Large Physics Course: Meaningful Patterns for Instructors. *Journal of Research in Science Teaching* 2003; **40**:776-791

3   Marbach-Ad G, Sokolove, P.G.: Can Undergraduate Biology Students Learn to Ask Higher Level Questions? *Journal of Research in Science Teaching* 2000; **37**:854-870

4   Hofstein A, Navon, O., Kipnis, M., Mamlok-Naaman, R.: Developing Students' Ability to Ask More and Better Questions Resulting from Inquiry-Type Chemistry Laboratories. *Journal of Research in Science Teaching* 2005; **42**:791–806

5   Gautier C, Solomon, R.: A Preliminary Study of Students' Asking Quantitative Scientific Questions for Inquiry-Based Climate Model Experiments. *Journal of Geoscience Education* 2005; **53**:432-443

6   Bransford J, National Research Council (U.S.). Committee on Developments in the Science of Learning. and National Research Council (U.S.). Committee on Learning Research and Educational Practice.: *How people learn : brain, mind, experience, and school*, Washington, D.C., National Academy Press, 2000.

7   Philip Bell BL, Andrew W. Shouse, and Michael A. Feder (ed)^(eds): *Book Learning Science in Informal Environments; People, Places, and Pursuits* Washington, D.C., THE NATIONAL ACADEMIES PRESS, 2009

8   Knapp D: Memorable Experiences of a Science Field Trip. *School Science and Mathematics* 2000; **100**:65-72

9   Fraser BJ: *Test of Science-Related Attitudes Handbook*, Hawthorn, Victoria, Australian Council for Educational Research, 1981.

Mater. Res. Soc. Symp. Proc. Vol. 1233 © 2010 Materials Research Society   1233-PP01-01

Marni Goldman Tribute: Contributions to Materials Science Education

Charles G. Wade[1] and Curt Frank[2]

[1]IBM Almaden Research Center, 650 Harry Road, San Jose, CA 95120

[2]Dept. of Chemical Engineering, Stanford University, Stanford, CA 94305-5025

Marni Goldman
1969-2007

Obituary

Marni was born with a severe form of muscular dystrophy that caused her doctors to predict she would not live beyond the age of two. She exceeded that prediction by 35 years, as a result of her amazing family and her own keen intelligence, indomitable spirit, high energy, tenacity, and love of life. Although she spent her life as a quadriplegic, Marni was able to accomplish much with her life. She earned two bachelors degrees from the University of Pennsylvania in Materials Science and Psychology and went on to earn a PhD in Materials Science from the University of California at Berkeley. In addition to her academic and professional achievements, Marni lived to the fullest in her personal life. She was able to independently drive a specially modified car through the help of technology and the reach of her multi-inch fingernails, always decorated in beautiful colors and designs. Marni surrounded herself with an amazing community of people who together shared books, games, theater, music, food, travel and so much more. And those who knew her are aware that chocolate was high on her list of rewards.

Marni was among the very first people to be hired at Stanford specifically for the purpose of designing and carrying out "science outreach" programs. She started her Stanford career in 2000 as Research Associate in Stanford's Center for Polymer Interfaces and Macromolecular Assemblies, under the direction of Chemical Engineering Professor Curtis Frank. She worked closely on educational outreach with Chuck Wade at IBM, a co-director of CPIMA. Marni later added to her tasks simultaneously jobs as the outreach coordinator for Stanford's Nanofabrication Facility and the Associate Director of Stanford's Office of Science Outreach.

Marni cared passionately about attracting students into science and engineering, with a special emphasis on those students who would increase the diversity of these fields by virtue of being an under-represented minority, a woman, or a person with physical disabilities. She directed programs that brought undergraduates from other universities to Stanford during the summers for research experience with Stanford faculty. She helped to create Stanford's Summer Program for High School Science Teachers that brought local teachers to work as interns in Stanford faculty members' labs. She worked tirelessly to help local middle school students in underserved areas on their science fair projects, and she helped to place disadvantaged high school students in internship positions in Stanford labs. She created an annual program that brought community college students – primarily under-represented ethnic minorities -- from throughout California to Stanford for day-long tours of Stanford labs and talks with Stanford faculty and students about science and engineering. Marni also was instrumental in creating a sense of community throughout the University among all of the faculty, staff, and graduate students engaged in various kinds of "science outreach" activities.

Marni was an amazing individual who touched the lives of all people who had the privilege of knowing her. She was an inspiration to all.

Curt Frank
Charles G. Wade

I'm one of the lucky ones-I got to work closely with Marni Goldman from 2000 until her death in February 2007. I marveled at her skills and determination from my first interaction with her (an impromptu interview at the MRS spring meeting in San Francisco) until our last phone conversation (just before the family vacation during which she died). The interview was for the role of education outreach director of the NSF Materials Research Center on Polymer Interfaces and Macromolecular Assemblies (CPIMA) at Stanford. CPIMA is an academic/industrial collaboration, the only one of the nearly 30 NSF MRSECs with an industrial lab as a true partner. Curt Frank at Stanford is the P.I. of this grant, and the organization of the center is such that the industrial co-director (me) has responsibility for educational outreach while the academic members have responsibility for industrial outreach.

Thus in our first encounter I interviewed this wisp of a woman sitting in a wheelchair and armed with an impressive resume, a soft voice, and probing eyes. The interview took place in a hotel near the Moscone Center in San Francisco. Since there were four sites in the center (UC Berkeley and UC Davis being the other two) I mentioned at one point the need to visit those, and asked how she'd come to the conference. "I drove from Berkeley." In what? "A van modified for my use." Where did you park? "On the street." The driving that she did around Berkeley was impressive enough but to fight traffic in mid-day in San Francisco and park at a hotel near the Moscone center - now that was a feat that would challenge the most fearless driver. I mentioned at one point that a primary requirement was organizational skills since the position required supervision of several programs, with many locations, many grants, and many deadlines. "If I didn't have superior organizational skills, I wouldn't be alive today". In short order Curt Frank, Brenda Waller (at the time the current educational director) and I hired Marni Goldman as education director, and my respect for her continued to grow.

The program was already successful. It had been in existence several years under other very able coordinators, but Marni built improvements over the next 6 years. She expanded the program with additional grants, formed a partnership with the Exploratorium Museum in San Francisco (for teachers), aggressively pursued Research Experience for Teachers grants, became active in national organizations supporting educational outreach, worked to increase minority participation in science with local high schools and community colleges, and established a niche program to encourage students with disabilities to participate in internships. Her responsibilities increased at Stanford, including involvement in the NSF Center for Nanotechnology Instrumentation and with the Stanford Nanofabrication Facility. She soon assumed a part time position in the office of Science outreach in the office of the Vice President for Research. In short, Marni was a dynamo and an impact player.

Over this time our interaction remained dominated by the student undergraduate research experience (SURE) program which brought 25 undergraduate students from across the US to do research at Stanford, UC Berkeley, UC Davis, MPI Mainz (Germany), and IBM. Selecting and assigning the students was a heavy load. Annually each of us would read and rate every one of the 160 to 180 applications, then meet to select the participants and match up projects. Marni then took over: contacting all the students, making changes in the assignments as necessary, arranging housing and stipend payments, disbursing travel money, planning the summer programs and arranging the other summer activities. At the annual CPIMA technical meetings she organized a poster session with perhaps 80 presenters (summer students participated along with graduate students and postdoctoral fellows). Marni became a thorough professional in all this. She had strong opinions about the purpose of the program and took it very seriously. Her goals were to encourage the participation in science of women, of members of underrepresented groups, of the disabled, and of students from small colleges. All of these groups benefited from her efforts.

Marni's success evolved from fierce drive and determination, and she could be a tough lady. For example, she insisted that any application for the summer program had to be complete before it would be considered. When I protested that the students shouldn't be penalized if one of the three professors asked to provide reference letters were delinquent, she disagreed. Her point was that part of the process was to get *all* that in order, and the students had to have, or should develop, the responsibility to see that the letters arrived. She expected them also to live by the rules she established for the program. A running joke developed between us that I was a softie since I always supported the students in our discussions.

In the years I worked with Marni we experienced many successes. Because of our close interaction, I was also afforded a long term view of how she reacted when life wasn't so successful. She certainly experienced down times, from setbacks in the program funding to automobile accidents to periods of hospitalization. One of the more remarkable talents that Marni possessed was the ability to keep such setbacks in perspective. Nothing seemed to upset her; she took all those problems in stride and dealt with them. She would express irritation when frustrated but usually it was accompanied by a chuckle. I cannot recall a single time in those six years when she complained or was in a bad mood or was irritable or was depressed. This is a special memory of Marni that I will always carry—that remarkably stable personality. It had a big impact on me and on others she encountered. There was no way one could complain when you considered what she dealt with on a daily basis.

Marni had a very active life, many friends, and many interests. She had an apparently inexhaustible supply of colleagues from the technical community whom she recruited for the career day sessions or other activities for CPIMA. Her family was very active and supportive, and they came up frequently in both action and conversation. Chocolate was her only admitted vice, and she arranged to have it at many sessions. Marni kept her nails long and multi-colored. They created an immediate visual impact and clattered as she did tasks. She told me once that she kept them long because they provided a discussion item for people who were hesitant to approach her because she was in a wheelchair. Her nails were often the first topic of discussion. She used these with great success in handling not only people but also pushbuttons, even ones as crucial as those in the computer interface with which she piloted her van. Finally, she drove the fastest wheelchair on campus and keeping up with her was an effort.

Indeed a most remarkable woman.

Charles G. Wade

Mater. Res. Soc. Symp. Proc. Vol. 1233 © 2010 Materials Research Society 1233-PP01-05

Addressing Diversity in STEM Education: Authenticity and Integration

Fiona M. Goodchild and Maria O. Aguirre

California Nanosystems Institute, University of California, Santa Barbara, CA 93106-6105, U.S.A.

ABSTRACT

This talk will reflect on the challenges of designing educational opportunities that broaden diversity in the ranks of future scientists and engineers. The speaker, who is Education Director at the California Nanosystems Institute (CNSI) at the University of California, Santa Barbara, will report on the design and evaluation of a program that integrates academic, career and social components to engage a community of undergraduates, graduate mentors and research faculty at UCSB. The program builds on key practices such as academic mentorship, community networking and early undergraduate research. Evaluation of this program, Expanding Pathways to Science, Engineering and Mathematics (EPSEM) indicates that it has been successful in recruiting and retaining students from under-represented (URM) groups into science, technology, engineering and math disciplines (STEM disciplines).

INTRODUCTION

Long-term investment in diversifying the STEM academic and professional ranks has had mixed results at all levels of the educational system. Female enrollment in higher education increased to the point that in 2001 women earned over half of the undergraduate degrees in science and engineering (Lowell and Hartzman, 2007). Women have increased their participation in STEM Ph.D. programs, reaching parity level (49%) in the Biological Sciences, but are still less than 20% of successful candidates in Engineering disciplines (NSF, 2005). Though this improvement is not yet reflected in the number of women at the faculty level, there is movement from 7.1% of faculty at Carnegie Research Universities in 1973 to 29.1%, thirty years later in 2003 (NSF, 2006).

In the case of underrepresented students, the picture is less promising. Current data reinforces that the gap in immediate postsecondary enrollment rates between high school graduates from high- and low-income families has persisted from 1973 to 2001 (2007). Approximate figures for undergraduate (15%), masters (11%), or doctorate level (7%) indicate room for improvement to reach the parity level of 32% for 18-24 year-olds. Moreover, URM faculty represent only 6.8% of Science and Engineering faculty at Carnegie Research Universities (2009, CPST).

Recent reports (NSF, 2008) acknowledge that it has been difficult to evaluate the impact of the investment in broadening diversity. Programs that target specific URM populations have not been well documented to track subsequent progress of participants. Current preference is to design projects that integrate a wider spectrum of students and set more realistic goals for student achievement and summative evaluation (NSF, 2005). This paper presents an alternative model for broadening diversity at the undergraduate level, building on the results of programs that stress the value of academic mentorship, community networking, early undergraduate research.

EXPANDING PATHWAYS IN SCIENCE, ENGINEERING AND MATHEMATICS (EPSEM) PROGRAM

Expanding Pathways in Science, Engineering and Math (EPSEM) includes several elements that are integrated with undergraduate academic courses and extracurricular activities. EPSEM integrates students from different backgrounds in a regional program that serves a range of students, some disadvantaged and underrepresented, others from traditionally well prepared backgrounds.

Summer - Residential Summer Institute in Mathematics and Science (SIMS)

SIMS grew out of an orientation program for that targeted low income and under-represented minority students entering first year at a major research university. That program focused on providing information about campus advising and resources. After tracking the progress of students in the first tow cohorts, it become clear that the program was not making a positive difference to the persistence and retention of undergraduates in STEM disciplines. At that point, science educators and faculty at CNSI proposed a new model that would select from a broader pool of student and would embed the bridge program activies in the realm of academic science, engineering and mathematics.

Recruitment: To widen the range of participating students, EPSEM advertised to all entering STEM students and selected applicants from different socioeconomic and ethnic backgrounds, reflecting a belief that the students needed to work with peers who bring a variety of expectations and preparation, and to develop more realistic expectations. Selection priority is non- traditional students, first generation college- bound, and those who have shown a pattern of seeking out and participating in academically enriching opportunities specifically in the sciences, engineering, and mathematics. The resulting mix of applicants and participants reflects fewer URM students after 2005 but maintains a high level of diversity, as shown in Table 1.

Table 1. SIMS Program Demographics on Applicants and Interns

	2005	2006	2007	2008	2009	Cumulative
# APPLICANTS	39	63	45	94	123	364
female	18	19	19	43	53	152
	46%	30%	42%	46%	43%	42%
minority	30	31	17	25	24	127
	77%	49%	38%	27%	20%	35%
# INTERNS	29	28	27	28	33	145
female	12	14	14	14	17	71
	41%	50%	52%	50%	52%	49%
minority	23	14	11	13	11	72
	79%	50%	41%	46%	33%	50%

Program structure: SIMS design was modified by including a group laboratory project so that students would get a chance to participate in authentic research and to interact with graduate researchers and faculty before they started their first year academic courses. While the summer residential program still includes more traditional information about academic support and financial aid, the major new focus is to arrange that students work in teams to find out how research is conducted, how data is collected and how scientific results can be presented. This format reflects the assumption that while it is important for students to become familiar and comfortable in the academic environment, students need to become more aware of intellectual demands so that they understand how to become competitive in their STEM coursework. In summary, the SIMS bridge program can be described in three main categories:

(1) Research experiences, social interactions and network building within STEM community: Students work in small research groups with graduate student mentors, matched according to their interests in certain SEM fields. Mentors orient SIMS students to the diverse and collaborative research culture and community at UCSB, as well to the process of scientific inquiry, by providing authentic hands-on challenges that build skills and confidence. SIMS students also work in interdisciplinary teams of 2-4 students to collaborate on a project that requires mathematics, science, engineering, and multimedia. UCSB Faculty and staff supervise students during their projects to establish links that may develop into mentorship throughout the academic year. Social events take place daily in the afternoons and evenings with undergraduate program assistants, program staff and UCSB faculty. Informal late evening discussions are hosted by the onsite Faculty-in-residence.

(2) Career Exploration: A seminar series targets the breadth of career possibilities available to STEM graduates. Faculty and industry visitors make presentations on different career pathways, while graduate students give talks about their previous research and scholastic experiences. A variety of lab tours emphasize the range of opportunities at a research university. UCSB faculty and their research groups host these tours to highlight and showcase their recent progress along with sophisticated technical equipment to excite and motivate the SIMS students. Field trips to high-tech companies such as the Jet Propulsion Lab, Raytheon, and Asylum complement the visits to the campus laboratories.

(3) Academic Skills: SIMS students take an introduction to university Calculus, Chemistry, and Technical Writing. Additionally, students participate in computer lab workshops, and academic strategy and career workshops on topics such as résumé writing, library search skills, stress and time management skills, and test preparation to prepare them their critical first year. SIMS staff and faculty are expected to try to identify and respond to each student's personal needs.

Fall - Community Building:

After completing SIMS, the first year students expand their social network from the cohort of 30 SIMS colleagues and undergraduate peers who had served as residential assistants, to the EPSEM community that includes about 90 currently enrolled undergraduate students and many graduate and faculty mentors. They engage in year-round activities within a support network that encourages them to keep working on the knowledge and professional skills necessary to complete their bachelor's degrees and to prepare to enter graduate study and the professional workforce. EPSEM students who advance within the program can take leadership roles, for example as mentors for entering participants or as office holders for the Society for the Advancement of Chicanos and Native Americans in Science (SACNAS).

Leadership: EPSEM cultivates student peer mentorship and leadership by identifying students who attain exceptional academic success, participate in summer programs and gain highest regard from peers and program leaders. These student leaders take on responsibility for running program activities and supervising less-advanced students. By cultivating leaders at all levels of the program, we build community and a sense of belonging for students through peer mentors who share their cultural and social diversity. The UCSB Chapter of SACNAS meets bi-monthly, bringing speakers to address relevant academic and social concerns and introducing researchers who inspire interest in new discoveries. SACNAS also hosts student social events that are geared to complement busy schedules of coursework and employment.

Winter - Early Undergraduate Research Experience and Knowledge Acquisition (EUREKA)

EPSEM students with a sound start in their introductory courses are encouraged to think about applying for a research internship in their second quarter. Many of them need to earn money to pay for their tuition and find that it is a tremendous advantage to do that in an academic setting where they learn academic content and practical scientific skills. The EPSEM coordinator requests project proposals from UCSB faculty, matches them with EPSEM student ability and interests and monitors their subsequent participation.

An important option for students at this point is a credit course that introduces them to the world of laboratory experimentation and discovery, entitled *"The Practice of Science."* This course is open to all undergraduate students at UCSB and is designed to open doors to the research enterprise in general and the profile of research at UCSB in particular. Sponsored by the Physics Department in conjunction with CNSI, this two-quarter undergraduate course is aimed at helping students develop potential careers spanning experimental science and technology, based on information typically absent from traditional coursework. Lab tours often complement the talks given in the course by guest researchers from UCSB and from the local community.

The course is team taught by CNSI faculty and consists of lectures, scientific presentations, and invited talks by research scientists from several departments and local industry to provide a broad and balanced perspective. Lectures on laboratory techniques, advanced instrumentation, data collection and analysis, and ethics of research are accompanied by practical exercises, laboratory visits, campus facility tours, and state-of-the-art research projects. This course is an opportunity to gain an appreciation and understanding of experimental science and engineering, and its importance in a scientific career. Guest speakers also talk about intellectual property and start-up companies, including the trials and tribulations of the effort.

In particular, students:
- Obtain an overview of the full scientific process, from inception to final outcome
- Interview faculty and graduate students to explore career opportunities. Write an account of these interviews that reflects student interest and experience
- Articulate concepts, create research proposals, and explore funding sources
- Plan a research agenda to strategically attack scientific and technical problems using all available resources on campus in nanoscience and technology
- Choose and execute specific research projects alongside graduate students
- Develop and establish collaborations and effective team efforts
- Prepare and referee short scientific publications
- Practice and deliver effective technical presentations
- Present a proposal for individual research to the class and any relevant graduate students and faculty

The student laboratory projects are the focus of the spring quarter, when classes are run in the style of research group meetings with reports and discussion of issues such as scientific instrumentation, data collection and the ethics of experimental research. Students are required to analyze data, draw conclusions, and defend their interpretations. This course has grown in popularity since it started in 2004 and the number of EPSEM students has risen from 4-5 in the first two years until 10-12 students 2008-2009.

EVALUATION

Results of Demographics and Tracking, 2005 to 2009

Diversity demographics of EPSEM applicants and interns for Summers 2005 (before the deisgn change) through 2008, as shown in Table 1, demonstrate our success at recruiting both female and minority students. Current national statistics indicate that only 15% of all STEM bachelor's degrees go towards minority students. The 55% minority participation rate in EPSEM greatly exceeds that.

EPSEM Program Impacts and Student Accomplishments

We have performed extensive tracking of the activities of former participants, from Summers 2005 to 2008. There have been a total of 112 students from these time periods. Of the 112 EPSEM participants from summers 2005-2008, 90 (80%) are still enrolled at UCSB and 84% of those are still in STEM majors.

Table 2. Academic achievements and retention of EPSEM participants

Year	# EPSEM Participants	Avg GPA EPSEM participants	Avg GPA for all students	% EPSEM Students still enrolled at UCSB	% EPSEM Students still in STEM	% Participating in Undergraduate Research
2005	29	2.78	3.01	69	70	30
2006	27	3.10	2.99	70	68	47
2007	28	3.23	3.02	89	100	40
2008	28	3.38	2.93	100	100	39

Student Publications and Presentations

Two EPSEM students have published their results; one in Chemical Physics Letters and the other in the Journal of Physical Chemistry. All students in the 2005 cohort who have participated in research have presented their work at local and national conferences. Two of them won awards for their work at the California Alliance for Minority Participation (CAMP) statewide symposium at UC Irvine. All in the 2006-2007 cohorts have presented at either a local or National Conference. Three won awards for their work at the National Society for Advancement of Chicanos and Native Americans in Science (SACNAS) in October 2008.

Student Leadership

31 out of 112 students have taken on leadership roles in science education programs, acting as peer-mentors in the EPSEM/SIMS Programs, teaching science to K12 students, or acting as chapter officers for UCSB-SACNAS which was recognized as the 2008 National Chapter of the Year.

CONCLUSIONS

The academic authenticity, faculty engagement and integration of a wide range of students have contributed to achieving such impressive results. The staff professional that coordinates the EPSEM program plays a key role, both in terms of recruitment as well as continuity and monitoring. She ensures that students in the EPSEM program receive individualized focused assistance at critical stages of their undergraduate study, beginning the summer before their freshman year and continuing through enrollment and graduation from UCSB. As a community the EPSEM students engage in professional development that increases their persistence and raises their eligibility for STEM professional opportunities.

ACKNOWLEDGMENTS

The authors acknowledge support from the California Nanosystems Institute at UCSB and from the National Science Foundation through a STEP award (DUE # 0336668).

REFERENCES

1. Commission on Professionals in Science and Technology (http://www.cpst.org)
2. Lowell, B. L., & Hartzman, H. (2007). Into the Eye of the Storm: Assessing the Evidence on Science and Engineering Education, Quality and Workforce Demand.
3. National Academy of Sciences, National Academy of Engineering, & Institute of Medicine. (2007). *Rising above the gathering storm: energizing and employing America for a brighter economic future.* Washington, DC: The National Academies Press.
4. National Science Foundation (2008). *Broadening Participation at the National Science Foundation : A Framework for Action.*
5. National Science Foundation. *Pathways to STEM Careers: Preparing the STEM Workforce of the 21st Century. Broadening Participation through a Comprehensive, Integrated System*: Final Workshop Report, January 2005. http://www.seas.gwu.edu/~stem/STEMreport_March05.pdf
6. National Science Foundation Science and Engineering Doctorate Awards (2005).
7. National Science Foundation Science and Engineering Indicators (2006).

Mater. Res. Soc. Symp. Proc. Vol. 1233 © 2010 Materials Research Society   1233-PP04-21

Identification, Development and Implementation of Nanoscience Activities for Alabama K-12

Martin G. Bakker[1,4], Katrina Staggemeier[1], Amy Grano[1], Aaron Kuntz[2], Jim Gleason[3], Leigh McKenzie[6], Brenda O'Neal[5] and Rachel Pace[7]

[1]Departments of Chemistry, [2]Educational Research, and [3]Mathematics, and [4]Center for Materials for Information Technology, The University of Alabama, Box 870336, Tuscaloosa, AL 35487
[5]Admiral Moorer Middle School, Eufaula, AL 36027, [6]Eufaula High School, Eufaula, AL 36027
[7]McWane Science Center, 200 19th St N., Birmingham, AL 35203

ABSTRACT

We report on a pair of MSP (Mathematics & Science Partnership) START pilot projects designed to identify nanoscience experiments that will fit within the Alabama course of study for use in Alabama K-12 classrooms. As part of the first project we are testing the development, refinement and evaluation of an activity already partly developed. The form of this activity has had input from a focus group of RETs who were tasked to provide input into the activity and how it can be matched to components of the Alabama Course of Study. This activity consists of using sparks generated by abrasion of misch metal by sand paper of different grit size. Different grit sizes produce metal particles of different sizes, resulting in sparks of different size and length. If done in a dry box no sparks are produced and the powder left is not pyrophoric, demonstrating that high surface area, heat and oxygen are all required to produce sparks. SEM characterization of the powder allows the particle sizes to be determined, giving the correlation between size, grit size and spark track length. The activity was tested on groups of middle school science campers at McWane Science Center, and after evaluation, further modified to increase student interest and impact. The activity was then tested on grades 6-8 in a middle school classroom by a graduate student/undergraduate student team.

INTRODUCTION
Alabama context

The impact of No Child Left Behind in Alabama includes rigorous adherence to the Alabama Course of Study which does not (currently) include Nanoscience or Nanotechnology. The level of funding for science education also means that few teachers have physical science backgrounds and that there is little funding for classroom resources. The state of Alabama does however provide internet and computer facilities, and supports science and mathematics through the Alabama Mathematics, Science and Technology Initiative (AMSTI) and Science In Motion (SIM) programs. The former provides sets of classroom sets of experiment kits which are sent out to participating schools. The latter provides single experiments including equipment such as gas chromatographs which are delivered to high schools. Both programs provide in-service training for teachers on the kits and equipment as well as back up support. However, neither program has a component focused primarily on nanoscience or nanotechnology.

MSP START

In order to increase the quantity and quality of nanoscience in K-12 Alabama classrooms, a grant to Tuskegee University provided support for partnership activities to develop components for an Mathematics and Science Partnership (MSP) proposal that would partner Alabama schools

45

with institutions of higher education, including research centers in nanoscience and nanotechnology. The University of Alabama has a Materials Research Science and Engineering Center which includes faculty active in nanoscience. Our team proposed two linked activities to identify, develop, test and implement one or more nanoscience themed experiments that would be suitable for use at middle and/or high school level. A complimentary task was to discuss what components should form part of the MSP grant to build viable, lasting school/higher education/research ties.

ACTIVITIES
Summer development
The core of the summer activity was weekly lunch focus group meetings of 5 RETs participating in an RET Site in chemistry. This group met with the supported graduate student, participating University faculty and master teachers from the regional AMSTI Site. Graduate student Ms.

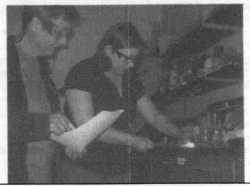

Figure 1 RETs Leigh McKenzie (left), Brenda O'Neal (right) and graduate student Katrina Staggemeier (center) striking sparks from sandpaper.

Katrina Staggemeier, searched the web for already developed activities, and presented these to our focus group for comments which resulted in the selection of two nanoscience themed activities for development, testing and implementation. Selection was based on perceived impact on the students, and how well the activity would teach nanoscience. One was based on an existing activity developed by the NISENet, (NISENet is the website of the Nanoscale Informal Science Education program) the other on an original idea. The latter activity called "Sparks", and subsequently "Nano-Nano" (from the 1970s TV program "Mork and Mindy") was based on the strikers used in chemistry laboratories. It starts with running a set of striker flints down a piece of sand paper to produce a shower of sparks (Figure 1), and then allows the students to make sparks themselves with different grades of sandpaper as part of an experiment to explore the effect of size, and friction on the sparks generated. This demonstrates the importance of surface area in controlling the rate of chemical reactions. The striking visual effect, and the hint of danger was anticipated to appeal to middle school students. Subsequent testing confirmed the safety of the experiment, although we insisted that safety glasses be worn. The second activity was an implementation of a NISENet experiment which similarly showed the effect of surface area using Alka-Seltzer tablets compared to crushed Alka-Seltzer. We also wanted to incorporate a mathematics component, and so an activity for the Sparks experiment was developed designed to illustrate the idea of similar triangles by seeing how far the sparks fell as a function of height. For the Alka-Seltzer experiment the height of the foam formed on mixing an Alka-Seltzer tablet with water was measured.

Summer Testing

Both experiments were tested at the McWane Science Center on the 23rd of July. This test was with two groups of science campers. The larger one was twenty-four 5th and 6th grade students. The smaller one was six 7th grade students. One group was faculty lead the other graduate student lead. Observations made by the team as well as by the independent evaluator included:

(1) That the experiments were successful in engaging the students, leading to significant student/team member interactions concerning the scientific content. Indeed, the "Sparks" experiment was sufficiently engaging that we needed to bring spare striker flints.

(2) The mathematics component for the Sparks experiment although successful, did not relate to a meaningful scientific question.

(3) The mathematics component of the Alkaseltzer experiment did address a meaningful scientific question. Students can tell the difference!

(4) It is important to use "age appropriate" scientific language: the older students knew about combustion and so discussing the science in this context was more meaningful and leads to good scientific questions.

(5) That science campers at McWane Science Center while perhaps appropriate as a first test of an experiment, were not very representative of middle school science students in general.

It was clear from the testing at McWane that a different mathematics component was needed. Our focus group had also made it clear that for teachers the most valuable activities were ones that had multiple components that teachers could pick and choose from, as well as suitable teacher background materials to give them confidence in their understanding of the material. It was also clear that the actual activities would need to be extremely low cost, and/or should use materials available within the experimental kits available from AMSTI. We therefore developed two new components.

Further Development

(1) The first addressed student questions concerning the importance of friction vs. combustion. To establish that combustion was important we used a dry box to exclude air and found that indeed although metal particles were produced there we no sparks. In the absence of a portable dry box to take out to classrooms we shot a video clip that demonstrates the experiment.

(2) To incorporate a mathematics component we took SEM images of the metal particles produced from the three different grades of sand paper, and developed a work sheet to allow students to calculate and average "mean" particle size. To test the hypothesis that rougher sandpaper lead to smaller particles and smaller sparks.

This was the experiment that we took to one of our local middle schools and tested on 6th, 7th & 8th grade students. Again the basic activity was found to be very engaging for all classes. The mathematics component was found to be much better integrated, and a number of students expressed delight to find that their average gave repeating decimals which they had just covered in mathematics. We also took the opportunity to discuss with the students nanoscience and nanotechnology including environmental concerns, as well as explain why NSF was supporting our graduate students to work with their class.

In the course of working with a number of classes we observed differences that middle school teachers are probably already well aware of: for 6th grade class the Socratic method works extremely well, 6th grade students are delighted to make suggestions and advance hypotheses. By 8th grade students have become very aware of not making public mistakes and are much less willing to verbally comment. We also observed that most graduate students would benefit by a

Figure 2. SEM images of spark particles. (A) from course sandpaper and (B) from fine sandpaper. Note the very different scale bars at the bottom of the images. The presence of the m in mm tied to the course of study objective on metric measurements. The very different length scales is an opportunity to discuss the range of length scales and the length scales for nanoparticles and atoms.

little training on taking charge of a class and guiding them between the various components of the experiments. This was true even with the teacher present at the back of the room to keep control of the class.

One observation proved both very illuminating and contains the germ for the next component of the experiment. We observed that when asked to pick three particles to measure and average (Figure 2), the students almost invariably chose the three largest particles, which suggested to us that this was a wonderful opportunity to discuss observer bias and to develop a module on how particle sizing is done. Analysis of TEM images to obtain particle and grain size distributions is routinely done by metallurgists, and cell biologists among others carry out similar analyses. This suggests that by looking at the older, pre-digital imaging processing literature we should be able to come up with a suitable algorithm.

We have also carried out elemental analysis of the particles in the images and found that all the misch metal particles were converted to the oxide, consistent with a combustion reaction, but that some of the particles must have been grit from the sandpaper. When implemented this will add a further level to the activity.

Evaluation comments on the refined activity show progress overall: *"Based on observations of the "Sparks Experiment" at Echols Middle School it is evident that both the content and delivery of the experiment have been refined to more effectively facilitate student learning. Students showed interest in the experiment and, importantly, linked classroom activities to scientific processes, a key aspect of the experiment that extends beyond strict content learning"*

The progress in the pedagogy was particularly satisfying: *"... this version of the experiment benefited from refined pedagogical techniques. Most notably, the use of handouts as a pedagogical tool proved successful and was more integrated into the experiment facilitation than seen in the past. Students were engaged with the handouts and linked their measurements with the actual practice of scrapping the sandpaper and producing sparks. Further, the language used to communicate experiment processes and findings seemed age-appropriate and in-line with student experiences."*

However as with everything that is worth doing well there is always of room to improve:*"... it is suggested that facilitators brainstorm a list of key scientific terms that accompany the experiment and discuss which are appropriate given the age of student participants. In this way, facilitators can intentionally incorporate appropriate language into the lesson and check back in with students at the conclusion of the experiment to gauge their learning and understanding of key concepts (these terms might also be listed in documents available to teachers who might incorporate the experiment into their own classrooms)*

At the beginning of the Fall semester we were given the opportunity to work with the Advanced Placement Chemistry class of one of the local high schools. After a series of e-mails with the teacher of this class, we came to the conclusion that neither activity we had developed was appropriate for this grade, and so we went back to the literature and the internet to identify a experiment we felt was suitable. We located in the literature an experiment on the synthesis of gold nanoparticles, which on testing appeared to be suitable for implementation and further development. Discussing the experiment with colleagues lead to sufficient interest that the experiment is likely to be taken up as part of a biochemistry flavored physical chemistry laboratory course for undergraduates taking our pre-health professions chemistry degree. However, when the time came to set a date for the activity with the high school repeated e-mails and phone calls went without response. We are now planning on taking this activity to Eufaula High School.

RESULTS
One of the outputs from our project is the new nanoscience activity that we have developed. The worksheets for the most recent version of the sparks experiment as well as the video clip of the sparks not being struck in a dry box have been submitted to the Alabama Learning Exchange (ALEX: http://alex.state.al.us/index.php) for web dissemination. We have also presented a workshop for 24 teachers on the activity at the recent Alabama Science Teachers Association conference, which was well received. Development of the further components of the activity, along the lines discussed above is on-going.

The other result of the project is at the programmatic level. Developing and testing a nanoscience activity was envisaged as a mechanism to prototype activities for an MSP proposal. As such our group was tasked to address how to build successful partnerships. As researchers, we developed a much better appreciation of the skills required to teach middle school students, and to teach them well. Our experience points strongly to the need to work co-operatively with teachers who have much greater familiarity with the state course of study. Conversely, we also confirmed the need for higher education partners to provide the content knowledge in this area. The extent of testing and development required to develop the activity to a point where it can considered to be working "well" was significantly larger than we had anticipated. This may reflect the need for greater teacher involvement during the testing phase. We were fortunate to be able to use an existing linkage with a local middle school, but in testing the activity we have exhausted our pool of accessible local middle school students. Outreach to other area schools and the two local school boards has not yet been fruitful despite a range of attempts and efforts. As a result, our next testing cycle is going to involve an overnight trip to Lake Eufaula to work with our RETs there. This set of classroom visits aims to do detailed pre/post test evaluation of student attitude and learning to provide a sound summative evaluation for our efforts. That we will be forced to drive 5 hours to obtain this data underlines the importance of building strong personal relationships. Our experience strongly suggests that lasting partnerships can not be built by "one-off" activities, but require a series of regular visits to do in-class activities to build strong partnerships. It also underlines the value of having teachers work in research labs as a way of building partnerships.

CONCLUSIONS

We believe that we have successfully developed, tested and implemented a new nanoscience experiment suitable for middle school classes, although appropriate summative evaluation is still to be carried out. Among lessons learned was the importance of developing a suite of related activities and supplementary materials, of the on-going nature of such activity development, and of the need to work closely with teachers all the way through the development cycle. The development of this activity served as the vehicle for refining a model for building Higher Education/school partnerships which we believe will be appropriate within our local school/university context.

ACKNOWLEDGMENTS

Primary support for this work by NSF MSP-0832129 is acknowledged, as is support by the Graduate School and the Central Analytical Facility at The University of Alabama.

Mater. Res. Soc. Symp. Proc. Vol. 1233 © 2010 Materials Research Society   1233-PP11-05

William R. Heffner[1], Himanshu Jain[1], Steve Martin[2], Kathleen Richardson[3], and Eric Skaar[3]

[1] Int. Mat. Inst. for New Functionality in Glass, Lehigh University, Bethlehem, PA 18015, USA
[2] Materials Science and Engineering, Iowa State University, Ames, IA 50011, USA
[3] Materials Science and Engineering, Clemson University, Clemson, SC 29634, USA

Abstract

As engineering becomes more and more specialized, both the faculty resources and number of interested students become limited. Consequently, very frequently highly specialized graduate courses are not offered, especially in disciplines like Materials with small faculty and enrollment. NSF's International Materials Institute for New Functionality in Glass (IMI-NFG) has successfully addressed this problem by successfully introducing the concept of multi-institution team teaching (MITT). It brings together via internet both the expert professors and students from many universities. By pooling the talent of various instructors, the courses become technically stronger and students learn advanced topics that would be available otherwise. As an example, a recent MITT course included instructors from 10 US institutions, and students from many more US and international universities.

Software such as 'Adobe Connect' is used for the live delivery of lectures, wherein students can see the instructor and Power Point slides as in a normal classroom. The students may ask questions any time during the lecture, and the instructor would respond immediately. They register and pay tuition at their home institution, so that no exchange of funds is involved between universities. Survey results support that a majority of the enrolled students liked the format and delivery of the course, and more than 75% students felt that multiple instructors, who "taught information of their expertise", made the course stronger. In conclusion, the concept of MITT has been successfully demonstrated for teaching highly specialized graduate courses.

1. Background

Glass science and engineering has been taught as a discipline of engineering for centuries, although only at very few universities. With increasing interest in more modern amorphous materials, many universities in the US hired faculty to teach glass in the late 20th century, while the traditional centers of glass education diversified into other materials. So even though the total number of professors at US universities, who are active in glass research may not have decreased over the years, most of such universities have just one token faculty member in glass science. This person typically teaches just one glass course, which ends up being an introduction to the whole field. Even when the lone professor of glass offers an advanced course, there are too few students in any given term, who sign up for such a specialized course, and it is difficult for the Administration to approve courses that have fewer than half a dozen students. The result is a larger number of students exposed to glassy materials, but with relatively shallow, cursory knowledge that does not prepare them to become a professional glass scientist or engineer.

NSF's International Materials Institute for New Functionality in Glass (IMI-NFG) undertook the challenge to correct this gradually but certainly deteriorating situation. It proposed to combine the resources of various educational institutions to share courses, by making use of remote teaching via satellite, Internet or a combination thereof. This approach has been tested successfully for the delivery of special education in rural areas [1,2] and nursing education [3], but not engineering education. So it began as a novel attempt on transferring specialized science and engineering knowledge to aspiring students rather than just through the traditional courses of lectures taught by one instructor in one classroom. To some degree, this basic idea was successfully tried at Lehigh University by a group called the Materials Pennsylvania Coalition (MatPAC), a network of six Pennsylvania-based universities with exceptional collective strength in advanced materials research and education. The MatPAC, comprised of Carnegie-Mellon University, Drexel University, Lehigh University, Penn State University, the University of Pennsylvania, and the University of Pittsburgh, had developed an educational model for multi-institutional sharing of courses via Internet2 to promote the field of advanced materials. In this case, a well established course taught by one professor is made available to students at all the member schools. As a result, each university is enabled to take advantage of the diverse expertise in materials education across the state. Two of the authors, under the sponsorship of IMI-NFG, proposed to explore the feasibility of cooperative teaching by glass professors at various US universities.

A course was then designed and offered in the Spring 2007 semester, followed by another course in Fall 2008. The student feedback indicated that the project was very successful. It was felt that the Multi-Institute Team Teaching (MITT) model could be a solution to the teaching of highly specialized advanced topics from many other fields of science and engineering where the enrollment or faculty expertise is too limited at any one institution. We hope that this report of our experiment on cooperative teaching and learning will help others to apply it in their areas of engineering as well.

2. Course Organization

The two cooperative MITT courses were intended to be second-level courses in glass on an advanced senior or elementary graduate level. They were organized to first review the material which would be covered in an introductory glass course. The review segment was covered in the first week of the semester. The remaining segments of the courses covered material new to the students and were taught by experts in their respective fields. Each segment emphasized a particular technique for structural characterization for the first course or an associated property of glass for the second course, using examples of the correlation between structure and properties of glass.

In the planning discussions among the instructors, structural characterization of glass was decided as the focus for the first cooperative course. This topic area was deemed to be fundamental and focused, building on the basics of glass science that each of the six institutions taught to their undergraduate cohort within an *Introduction to Glass Science* course. Thus the target student audience for the course was aimed at senior undergraduates who had taken this *Introductory* course, or first year graduate students who had only limited exposure to glass and glass formation processes. It was determined in a post course evaluation that this defined "starting point" level was suitable to most students taking the course.

There were several logistics problems related to scheduling and matching of calendars, not to mention the challenge of getting the course's approval from the Administrations of the participating universities. There was much excitement about the teaching of this course, which helped in resolving all the problems and overcoming initial hurdles. As a result, each institution allowed the course to be offered as an experimental course with the appropriate institutional identification number, with the local professor being responsible for its execution. There were no financial transactions involved in this teaching experiment. The classes were taught over the Internet as lectures, with the help of powerful software, *Macromedia Breeze (2007) and Adobe Connect (2008)*. It allowed each student to see the lecturer in one small window on computer screen, whereas the much larger second window showed PowerPoint slides (see Fig. 1 and Section 3). The students could ask questions any time during the lecture by typing in yet another window.

The first course was taught by six instructors from six different institutions spread across as many states in the East and Midwest. It was taught live with active communication between the instructor and students. The class met twice a week from 5:00 to 6:30 pm EST on Mondays and Wednesdays, making it convenient for graduate students to attend in two different time zones. In addition to the six universities represented by the instructors, the course was also taken by the students from Penn State University. The day-to-day operation of the course was managed from Clemson University. The lectures were also archived at Lehigh University at the IMI-NFG web site for distribution to the wider worldwide glass community. The recordings of the lectures were made available to the students for later viewing at their convenience and pace. It was particularly helpful to students when the links to archived lectures were available within a day or two following the lectures. Altogether 56 students and postdocs signed up for the course, out of which 28 took it for credit.

Each instructor tested the students on his or her part of the course via homework assignments, quizzes, or take home exams. A multi-institution team-based project was structured for the course's final examination. For this purpose, the students were divided into nine teams, each consisting of three or four students from different institutions, so that they had to collaborate over the internet and use other means of long distance communication. Each team was asked to analyze a different topic or issue of the structure of glass. The specific assignment was for the students to develop an experimental research proposal to solve specified glass science and/or technology problems. They were asked to prepare and present to the instructors and the rest of the class for review a short poster summary. The posters were electronically submitted to the course web page, and graded by each instructor separately. The students then presented their posters at the Spring meeting of the Glass and Optical Materials Division of the American Ceramic Society. There they met with most of their classmates for the first time and presented their proposal to a broader group of glass and materials professionals. By working with colleagues from another university and without physical face-to-face meetings, the students experienced how to collaborate with peers in an international Internet environment. Aside from some initial student resistance, the team project concept worked well, and by the end the students appeared to appreciate the experience. The grades on all the assignments and the final poster were made available to students and instructors via the course web page maintained on Clemson University's *Blackboard* course web page system. The final grades for the semester were then assigned by the local instructor according to the norms of his or her university.

Encouraged by the success of the first MITT course, the core faculty decided to offer another course, *'Physical Properties of Glass'*. In fact, there was such enthusiasm within the glass community that the second course was taught by nine glass faculty from as many institutions. We also expanded its delivery outside the United States with a very eager group of students from Brazil. Significantly, it turned out that one of the instructors, who happened to be on sabbatical leave in The Netherlands, delivered her lectures from there. This gave us an opportunity to test the international aspects of this course offering modality. The level of the targeted audience and the organization of this second MITT course were along the lines of the first course with one difference: each instructor gave a relatively more elaborate homework, almost like a take-home exam on the topic of his/her part of the course. Consequently, there was no final exam, and the course grade was based on nine take-home assignments.

3. Distance Learning Mode and Technology

Throughout the ages, teaching has gone from the apprentice style to the lecture style to now the distance lecture style. Each of these styles has advantages and disadvantages. Perhaps the most in-depth learning takes place under the apprentice style of teaching, but it is restricted to very few students in a single geographic location. The lecture style accommodates larger numbers of students, but they must still be in a single geographic location, the same location as the lecturer. Depending on the size of the lecture, more or less dialog can take place. Larger lectures accommodate fewer dialogs; nevertheless the lecturer can gage his or her effectiveness by general feedback from the audience. Distance presentations allow material to be taught to a large number of students in many different geographic areas. They can be synchronous where everyone attends the presentation at a single time or asynchronous (such as on-demand lectures) where the student can view the presentation at her / his convenience. The problem occurs with feedback from the students to the lecturer. Generally, the lecturer cannot see or hear the audience, and therefore cannot gage his or her effectiveness by general reaction or body language. In the case of asynchronous learning the lecturer has no real time clue at all.

Even though distance teaching eliminates much of the general feedback from the students, it nevertheless offers some advantages that make it very attractive for some situations. For example, in our case, we are interested in a topic which is rather specialized, and would not attract the numbers of students as would a class on a topic such as statics or dynamics. In this case, in cash strapped universities, the opportunity for teaching such a topic would be limited at best. By utilizing the distance techniques available today, such a class can be taught simultaneously on several campuses throughout the world by the top experts in the field. The students can be exposed to techniques, results and nuances generally not available from a local lecturer. Moreover, the expert can deliver his or her lecture from anywhere (s)he happens to be and the students can attend the lecture from any place convenient to them. The flexibility of distance lecturing opens up wide vistas of opportunity for education. Furthermore, from the university administration perspective, the effective faculty "cost" for one's effort in these courses is reduced significantly by the participation of external instructors.

In our cooperative glass courses, we utilized the synchronous approach, which required some scheduling compromises due to the time zones we covered. All of our students and most of our lecturers were in the western hemisphere, and as such we did not have anyone attending class at 2 AM or some other such inconvenient time.

Our lectures were hosted by the Clemson University servers and broadcast out over the internet. The only equipment required for the students was a personal computer equipped with a fast internet connection, modern browser and sound. Similarly, the instructors only required the student configuration with the addition of a microphone and camera. With modern laptop computers and high speed internet connections readily available, this mode allowed both lecturers and students to attend class while traveling, and indeed this occurred on several occasions. Additionally, each lecture was recorded, and those who missed the lecture were able to view it asynchronously.

In some universities the internet feed was projected in the classroom, and the students attended the lectures as a class with their local instructor, rather than individually. This paradigm works well, and students seem to like it. One distinct advantage in this case is that the local instructor can field some questions and refer others to the expert lecturer as well as expand upon the lecture material.

The servers hosting the class were running the *Adobe Connect®* software. A screen shot of a distance lecture is shown in Fig. 1. The main features of a lecture are: a motion picture of the lecturer, the class attendance, the chat box and the presentation area. Within the presentation area, the lecturer has a pointer and can, if desired, annotate slides using whiteboard electronic tools. Questions and feedback from the students are given in the chat box, and appear in real time to the instructor and the class participants. This particular feature is the main real time feedback the instructor gets from the class. However, we have found, in this era of instant messaging and texting that students do not hesitate to use the chat box to ask questions or give other feedback. It is actually a rather good method of getting feedback.

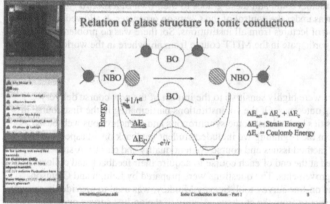

Figure 1. Screenshot of the cooperative glass course showing picture of lecturer, attendee list, chat area and presentation slide.

The final feature of our software is the motion picture of the lecturer. This particular feature has some advantages and some disadvantages. The main advantage to seeing the lecturer in real time is that the student wants to know or get some idea of the person who is talking to him or her. Additionally, the "talking head" distracts attention from any hesitations which always accompany a lecture. The disadvantage to this feature comes from the fact that a lot of data has to be transmitted for the motion. If internet conditions are particularly busy, this data can interfere with the audio or other data being sent over the internet. We found that in some cases we needed to freeze the motion for an acceptable lecture.

In general, our experience with these distance MITT courses has been favorable, with the advantages outweighing the disadvantages. We certainly intend to continue this form of course delivery.

Although technology has developed far enough to enable the transmission and reception of lectures remotely, there can be non-technical challenges in its implementation. Here we cite one hurdle and describe briefly how it was resolved. Since the first course was delivered smoothly from the 'control room' at Clemson University, which utilized *Macromedia Breeze* and *Blackboard* software systems, we decided to make no changes in this respect for the second course. Soon we realized that the delivery platform was upgraded to *Adobe Connect®* software. This modification was rather small as the faculty readily adapted to this change. However, we faced an unexpected challenge. The *Blackboard* system could not be used across the partner institutions, thus placing restriction on the distribution of course material, assignments, grades etc. Fortunately we had implemented the alternative access to course materials through the website of the sponsor, International Materials Institute for New Functionality in Glass (IMI-NFG). As the course continued, the Glass Properties webpage at IMI-NFG (www.lehigh.edu/imi) became the resource that the students could access for all the relevant information – including syllabus, lecture schedules, slides, homework, due date and even homework solutions. We ensured that all slides were available prior to the class. Once implemented, the manual collection and distribution of grades by the IMI-NFG staff served the purpose, and the host instructors could add these grades to their own *Blackboard* type systems.

The *Adobe Connect* was under the control of the Clemson organizer. It allowed remote transmission and reception of lectures from all institutions. So, there was no problem for anyone with access to internet to participate in the MITT course from anywhere in the world.

4. Student Evaluation

The course instructors were highly sensitive to the impact of the new course delivery method to student learning outcomes. Of the six institutions participating in the first course in 2007, most of the Universities had at least some synchronous and asynchronous web-based course offerings. Thus from the beginning we included student feedback on all aspects of the course including delivery method issues and course content quality and clarity. A student questionnaire was prepared at the end of each course to acquire their feedback and collect information for future improvements. The questions were prepared by Lehigh and Clemson University organizers. The on-line survey tool, Survey Monkey, was used to provide the survey and collect anonymous response from the students. A copy of the on-line survey questions used by the students for the most recent course is available on the IMI-NFG website at: http://www.lehigh.edu/imi/docs_GP/SurveyResultsShare.pdf

According to the survey taken after our first course, a three-fourths strong majority felt that the course was made stronger by bringing in multiple instructors, each teaching within his or her area of expertise. However, the feeling towards on-line format was mixed: 60% indicated that they liked the on-line format of the course. With regard to the level of course, 50% indicated that the course content was at the right level. There appeared to be somewhat of a mismatch between the level of lectures and students' background. It seemed that although we had succeeded in bringing the experts into the classrooms of multiple universities, there was a need to establish among the instructors a uniform level of expectations with regard to the students' background. Indeed the discrepancy was generally within the second course, for which we had more time to plan and a larger faculty pool. For this latter course, we took a more traditional lecture approach, and put extra effort on class organization and making material available to the students in a timely manner. Improved results from this latest course are described next in some detail.

The survey link for the Glass Properties course was sent to all students in the class (37), who submitted homework assignments, but did not include the "auditing" students, postdocs or faculty. Eighteen (49%) students responded, of which sixteen took the course for grade and the remaining two indicated their participation was not for grade. The survey was anonymous.

Overall two thirds of the students rated the course as good to excellent. A great majority (82%) of respondents found the level of the course just right, with one student finding the material too easy and one student finding it too hard. Most of the students (88%) found it beneficial to have multiple instructors, supporting one of our primary hypotheses. Also they found the *Adobe Connect* software interface to be a satisfactory vehicle for the delivery of this course. Only one student indicated disagreement with the Adobe Connect as a suitable learning format. Audio and video quality was fine for most (89%) students with 11% students finding it unsatisfactory, probably due to the issues of limited bandwidth.

Web page management for the course documents via IMI-NFG web site, including class lecture notes and associated materials, was considered satisfactory. Most students indicated that the webpage approach worked as well as or better than the conventional course management software such as *Blackboard*. Part of the success of our web page distribution solution appears to have been the prompt posting of all new lecture material, as confirmed by 94% respondents. The overwhelming majority of students (89%) found it useful to have the lecture slides available before the class and all students (100%) found it useful to have the archived videos of the lectures available to review again after the class. In fact, 61% indicated they would likely use the archived video lectures again in the future, after the course.

The students appreciated the opportunity to meet and learn from multiple faculty members from across the country. Two additional items that received multiple positive comments include: (i) the posting of slides ahead of class and (ii) having recorded lecture videos available shortly after each lecture for review.

Despite some room for future improvement, the student feedback suggests that the Glass Properties Course was an effective learning opportunity with a much higher student satisfaction than for our initial course. Thanks to the archiving, this course remains available for future use by all students.

5. Recommendations for Future Improvement

From student comments two issues are identified for future improvement: a) the discussion and feedback on homework, b) some students had trouble with the streaming technology, presumably due to limited local bandwidth.

By far the primary weakness of the course was identified with the homework. This was confirmed with a response of only 41% finding the homework useful and appropriate as well as many comments indicating complaints about the homework. The students' primary concerns were about the appropriateness of the assignments (some too difficult, too much time spent on looking up data and not always supporting the topics of the lecture) as well as too little opportunity for discussion and feedback from the instructors on the homework assignments. The latter also included concerns about: timely feedback on homework grades, availability of solutions, only getting graded assignments returned by less than half the instructors, and insufficient opportunity to discuss the homework assignments after their completion. Clearly, this is one aspect that should be addressed in future courses. One solution to this problem could be a greater engagement of the local faculty in all homework assignments, so that they can serve as *recitation instructor* for additional clarification and comment when needed by the students. The schedule should include a block of time for students to discuss issues with their local instructor or the distance lecturer after they have had some time to digest the material. As one student comment summarized, "Asking questions at the end of the lecture didn't work so well."

Internet data bandwidth was an issue for some of the students attending the course. We could see it during the lectures – some locations were experiencing trouble with delayed or intermittent audio while most other locations could hear the transmissions fine. Although most students did not seem to have any complaints on the audio/video quality and bandwidth, we received comments from two (out of 18) students about this issue. One student made an excellent suggestion that the IT department at the university should be made aware of such courses in advance, so that it can provide some priority service connections for the course. The students experiencing difficulties with bandwidth were not always aware that the problems were from their local systems; nevertheless, this was a frustration for some.

6. Concluding Remarks

It is not uncommon for an engineering faculty member to find that one of his or her courses cannot be offered due to insufficient student enrolment even though the subject is of high technological relevance. This appeared to be a pervasive problem in the field of our interest, namely glass science and engineering, due to its small size and the spread of experts at numerous institutions. We attempted to overcome this problem by introducing the concept of multi-institute team teaching (MITT). It combined advanced distance learning technology with unprecedented cooperation among faculty from several universities to share the teaching as well as their students. Our most recent course was taught live by professors from nine universities (Alfred, Arizona, California-Davis, Coe College, Florida, Iowa State, Lehigh, Michigan, and Missouri Sci. Tech.) and attended by over forty students, postdocs and faculty from many universities spanning five time zones across the continental United States and Brazil. Our experiment has demonstrated that logistics and course organization can be worked out to give students quality educational experience. Finally, for the success of MITT, it was important to start the planning significantly earlier than the usual course, to schedule an acceptable lecture time across multiple universities prior to registration, to rehearse the technology ahead of the lecture, and be open to innovation. In future, we plan to expand the concept of MITT gradually to overseas institutions from different regions of the globe.

Bibliography

1. J. Grisham-Brown, J.A. Knoll, B.C. Collins, 'Multi-university collaboration via distance learning to train rural special education teachers and related services personnel', J. Spec. Educ. Technol. 13 (1998) 110.
2. L.W. Olivet, T.C. Jones, 'Collaborative teaching: a strategy for interactive telecourses', Nurse Educ. 22 (1997) 6.
3. B.C. Collins, M.L. Hemmeter, J.W. Schuster, 'Using team teaching to deliver coursework via distance learning technology', Teach. Educ. Spec. Educ. 19 (1996) 49.

Mater. Res. Soc. Symp. Proc. Vol. 1233 © 2010 Materials Research Society   1233-PP04-12

Tackling Science Communication With REU Students:
A Formative Evaluation of a Collaborative Approach

Carol Lynn Alpert,[1,2] Eliot Levine,[3] Carol Barry,[4] Jacqueline Isaacs,[5] Alex Fiorentino,[1,2,] Kathryn Hollar,[5] Karine Thate[1,2]

[1] Museum of Science, Boston, MA, USA [2] Nanoscale Informal Science Education Network, Boston, MA, USA. [3] Donahue Institute, University of Massachusetts, Hadley, MA, USA [4] Center for High-rate Nanomanufacturing, University of Massachusetts-Lowell, MA, USA [5] Center for High-rate Nanomanufacturing, Northeastern University, Boston, MA, USA [6] Nanoscale Science and Engineering Center, Harvard University, Cambridge, MA, USA

ABSTRACT

This paper reports on a multi-faceted evaluation of Science Communication Workshops conducted during the summer of 2009 within the context of multi-week summer Research Experience for Undergraduate (REU) programs at four universities. The REU programs were associated with the Center for High-rate Nanomanufacturing (CHN) at Northeastern University, the University of Massachusetts-Lowell, and the University of New Hampshire, and with the School of Engineering and Applied Sciences (SEAS) at Harvard University. The design, content, and conduct of the science communication workshops resulted from a collaboration between faculty at these schools and staff from the Strategic Projects department of the Museum of Science, Boston. A formative evaluation of the Science Communication Workshops was conducted by the University of Massachusetts Donahue Institute's Research & Evaluation Group. The findings of the evaluation are being used both to improve succeeding iterations of these workshops and to ready them for dissemination as replicable models to other universities partnering with science museums, particularly those operating in association with the Nanoscale Informal Science Education Network (NSF ESI-0532536). Results of the formative evaluation indicated that the science communication workshops incorporated into the REU programs (1) substantially increased student interest in exploring and understanding the broader impacts of research and (2) substantially increased student knowledge, confidence and practice of communication skills for both professional and non-professional audiences.

INTRODUCTION

Many university science and engineering departments as well as research centers host National Science Foundation-funded "Research Experience for Undergraduate" (REU) programs, often during the summer months. As the NSF program solicitation [1] describes it:

> The Research Experiences for Undergraduates (REU) program supports active research participation by undergraduate students in any of the areas of research funded by the National Science Foundation. REU projects involve students in meaningful ways in ongoing research programs or in research projects specifically designed for the REU program.

REU students often work side-by-side in the lab or in the field with supervising researchers, such as faculty members, post-doctoral students and/or graduate students, taking on the investigation of one small but helpful aspect of the group's overall research program as their own summer project. REU programs often include some professional development activities and some social activities for participating students.

REU students are typically required to give a presentation on their research project shortly before the end of the research program. A Stanford Research Institute study of REU programs at NSF-funded engineering centers found that 86% of the students are expected to deliver a final research presentation [2]. Nevertheless, the SRI study reports, only 57% receive any training at all in science communication skills. Where included in REU programs, science communication training tends to prioritize the skills students need to acquire in order to write scholarly reports for publication. Programs that offer training in spoken communication are rare. And yet, spoken communication is not an insubstantial aspect of the professional lives of most contemporary university-based researchers. They teach, they mentor, they give presentations on their work to review panels and at conferences; they sometimes perform education outreach; they may find themselves with an opportunity to work with patent attorneys and venture capitalists. Those working in highly interdisciplinary teams, as in nanoscale science and engineering, require facility in cross-boundary communication with a broader community of researchers. While these reasons alone may be sufficient to persuade research faculty to include spoken communication skill training as a component in student preparation at all levels, there are additional reasons why doing so might be particularly appropriate at the undergraduate level.

Many undergraduates who participate in REU programs do so in order to help them decide whether to continue to pursue a career in research [2]. Social and family pressures may lean in other directions, particularly for women and minority students with fewer familiar role models. Undergraduate students may have difficulty communicating - to friends and family without science backgrounds - their own growing interest in science and even what they do in science classes and laboratories. Training in spoken science communication may provide not only an opportunity to enhance young researchers' academic and professional competence, but also their personal confidence and self-esteem, and their ability to better communicate what they do within these broader social circles.

Though this line of reasoning may seem quite obvious as a motivation for developing a spoken science communication program for REU students, it was not, in fact, the line of thinking that led to the development of our REU Science Communication Workshops.

PROGRAM BACKGROUND

Two National Science Foundation-funded Nanoscale Science and Engineering Centers (NSECs) have formed long-term partnerships with the Boston Museum of Science's (MOS) Strategic Projects team to further their education and outreach efforts and to address other "broader impacts," as mandated by the NSF [3]. These collaborative efforts have been supported by subawards from the NSF NSEC research grants to the Museum. Since 2001, the partners have produced dozens of nanoscale informal science education programs, workshops, exhibits, television news segments, and multimedia that reach a regional as well as global audience [4].

As the partnerships deepened over the decade, collaborating faculty and museum educators sought new ways to involve graduate students, postdoctoral students, and other "early-career" researchers in education outreach activities, including face-to-face interactions with public audiences at the Museum and K-12 students in neighborhood schools. The Museum staff wanted to be able to draw on these younger, generally more enthusiastic and more diverse pool of researchers in order to be able to better capture the attention and imagination of middle-school age students who begin in high school to track toward academic programs that lead to higher education and careers in science. For their part, the research center faculty were interested in giving their students opportunities to get involved in outreach and service activities [5]. Therefore the science communication workshops the MOS team initially designed for REU and graduate students associated with the Centers focused on developing research student competence in preparing talks and demonstrations for primarily public audiences.

Before long, a pedagogical conflict emerged, especially within the context of the REU Science Communication Workshops. Like most REU students, these students were engaged in preparing a research talk on their summer projects, and the expectation of faculty and supervisors was that the students would strive to deliver a research presentation at the level of professionalism that might be expected at a department colloquium or before a funding review panel. Such a professional presentation would be designed in a style very different from that one would use to deliver a public talk on one's research for a middle school or museum audience. During a 2007 REU Science Communication workshop in which one museum presenter advised an audience of REU students of Edward Tufte's advice [6] to avoid using PowerPoint presentations, and then informed them that if they had to be used, slide backgrounds should be black, and pictures should be substituted for words wherever possible – good advice for science museum presenters, but confusing advice for young scientists with PowerPoint-templated research presentations and hard-copy print-outs due to their NSF-funded advisors in a week's time – we realized we needed to revamp the approach.

We decided to work more closely with the senior faculty to better integrate the Museum science communication workshops within the arc of the entire summer REU agenda of professional development, research, and reporting. We would work together to help students achieve success according to faculty-prescribed professional standards and expectations in the context of science research and research reporting, while also encouraging students to seek a broader understanding of the context and meaning of their research and to practice techniques for explaining it to friends and family who might have little background in science. We would do what we could to support the intention of these students to consider a career in science by increasing their professional competence and their ability to share their work in larger social contexts; we would save the training in public speaking skills, inquiry-based learning and education outreach for later, when having made the choice to pursue a graduate career in science, they could begin to take on these further communication challenges. Those workshop elements remain as part of the team's Sharing Science graduate student workshops and Science Communication Internships for graduate students and postdocs.

PROGRAM DESIGN

The 2009 REU Science Communication Workshops (SCW) were developed as an integrated experience within a ten-week strand of skill-building seminars offered by the faculty program directors: Jackie Isaacs at Northeastern, Carol Barry at the University of Massachusetts-Lowell, Glen Miller at the University of New Hampshire, and Kathryn Hollar at Harvard University. Twenty-two students from the first three universities formed the CHN REU cohort and thirty-eight students from Harvard's School of Engineering and Applied Sciences formed the second REU cohort.

The SCW was split into two sessions for each cohort. The first session took place near the start of the summer program and the second session took place near the end of the program, shortly before student research presentations were to be delivered to faculty and peers. Each session ran for about four hours, and, in deference to student schedules, took place on a weekday afternoon and included lunch. The sessions included a mixture of talks, demonstrations, role-playing activities, writing exercises, and small group work.

Session One
- An introduction to science communication
- Dimensions of presentations and assessment criteria
- Voice, speech, and movement
- Presenting yourself and your work in brief, face-to-face situations
- Context and meaning in science communication
- Context and meaning exercises – both written and spoken
- Introduction to research presentation format and slide design
- Assignments to explore context and potential applications of REU student research projects, to observe and critique research talks, and to practice brief introductions. Return to second session prepared to deliver first half of final research presentation with slides.

Session Two
- Debrief assignments
- Warm-up and review: presenting yourself and your work
- Student delivery of first half of final research presentations with slides in small groups with faculty and peer feedback
- Debrief and discussion

These sessions were among the few times that REU students from the three participating CHN universities got together in the same space and interacted with one another. While the CHN REU program faculty had offered some sessions for students on literature searches, writing papers, and graphic design between the two SCW sessions, these were mostly conducted via videoconference. Thus the workshops also provided a social context, where students could interact face-to-face with other students from different backgrounds and different areas of investigation. This factor helped provide a suitable workshop atmosphere for practicing science communication with peers just outside one's own laboratory comfort zone.

Evaluation Procedure

The Center for High-rate Nanomanufacturing (CHN) contracted with the University of Massachusetts Donahue Institute (UDMI) to conduct an evaluation of the entire CHN REU program and within that overall evaluation, a more focused study of the Science Communication

Workshops, including the cohort of REU students from the School of Engineering and Applied Sciences at Harvard University.

Pre and post surveys were administered to each cohort during each of the two workshop sessions. These were anonymous surveys with a code that allowed the surveys from each session to be matched by individual. The Donahue Institute researcher, Eliot Levine, also conducted follow-up focus groups with the participating CHN REU students and administered a web-based survey to them after the end of the summer program. He also conducted confidential "stakeholder" interviews by telephone with faculty and staff from all the participating institutions. Fixed-response items were analyzed using standard quantitative and descriptive techniques, assisted by PASW 18 and Microsoft Excel. Open-ended responses were analyzed using a standard qualitative technique that involved multiple readings of the data and the assignment of themes around recurring ideas. Once themes were identified, each response was coded by its appropriate theme, and the patterns that emerged were described.

DATA AND FINDINGS

The Donahue Institute report [7] focused on the CHN cohort of students. Here we will present the data from that cohort and later comment on slight variations observed in the Harvard SEAS cohort. Many of the questions in the pre- and post-surveys focused on assessing the usefulness and relative value of particular components of each session. These are of significance chiefly to program staff who are seeking ways to improve succeeding iterations of the Science Communication Workshops. Here we will focus on data that present a picture of the perceived value and effectiveness overall of the SCW component of the REU program.

The CHN cohort of 22 students included 68% male and 32% female students. Seventy seven percent were Caucasian/White (N=17), 9% were Hispanic/Latino (N=2), 9% were African American/Black (N=2), and 5% were Asian (N=1). One student reported having a disability. Academic majors of REU students were mainly in engineering, physics, chemistry, and biotechnology.

Perceived Value of Science Communication Workshops

Overall, students valued the SCW program as one of the most valuable elements of their REU experience. In the post REU program survey, students were asked to compare the Science Communication Workshops to the summer's other REU training activities. On a scale from '1' as 'one of the least helpful' to '7' as 'one of the most helpful,' the average rating was 5.9, and no student gave a rating lower than 4. On a scale from '1' as 'one of the least enjoyable' to '7' as 'one of the most enjoyable,' the average rating was 6.0 and no student gave a rating lower than 4. [See Figure 1 and 2.]

Figure 1

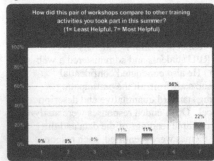

Figure 2

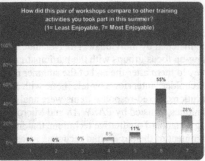

Session One Results

The Session One post-survey indicated that the most useful aspects of that session for students were the introduction to science communication, the discussion and exercise on discovering the larger context and meaning of their research, and the "introducing yourself and your work" exercises. The impact of these Session One elements was further clarified by data from the Session Two pre-survey.

The Session Two pre-survey asked students to reflect on Session One, their work between the two sessions, and their preparedness for the research presentation they were scheduled to make during Session Two. All respondents agreed or strongly agreed that Session One had "significantly increased their interest in seeking out and understanding the broader impact of their own and others' research [Figure 3]." This influence emerged as the strongest of all the impacts of Session One.

Figure 3

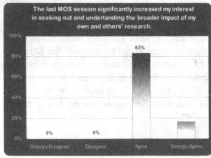

In addition, 78% of the students agreed or strongly agreed that Session One had significantly increased their confidence in "verbally introducing myself and my research in a variety of casual situations" [Figure 4], and 94% agreed or strongly agreed that they were "very pleased to have the opportunity to practice their research presentation" during the upcoming session [Figure 5].

Figure 4   Figure 5

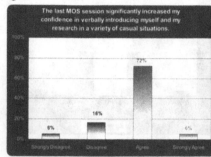 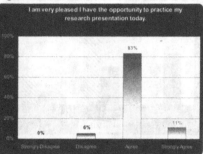

Session Two Results

The Session Two post-survey repeated several of the questions included in the Session One pre-survey as a means of noting change in perceptions over the course of the summer. Across the two sessions, student agreement that they had a good understanding of how to present their research to scientific audiences increased from 50% to 94% [Figure 6], and student agreement that they had a good understanding of how to present their research to non-scientific audiences increased from 50% to 100% [Figure 7]. Student agreement that they had a clear understanding of "one or more potential applications" of their summer research increased from 81% to 100% [Figure 8]. While it is impossible to isolate the effect of the Science Communication Workshop sessions from other experiences the REU students had over the course of the summer program, the SCW sessions were the only activities dedicated to these learning objectives.

Figure 6

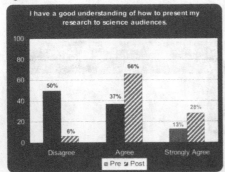

Figure 7

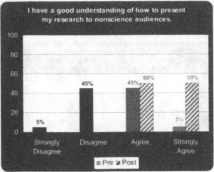

Figure 8

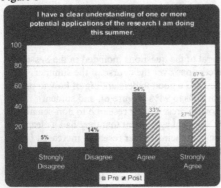

Web-Based Post Survey Results

This survey was administered after the students had completed delivering their final research presentations and the REU program had ended. All 22 students responded.

Ninety-one percent of the students reported that their ability "to understand how a particular science or engineering challenge relates to a larger goal or application" increased a lot or a little, and 95% reported that that their ability "to construct a professional PowerPoint presentation" increased a lot or a little. Ninety-one percent reported that their ability "to communicate [their] research projects and results verbally as a 15-minute presentation" increased a lot or a little, and 95% reported that their ability "to summarize the purpose and results of a research project in a brief 1-3 minute 'elevator speech' to other researchers in the same field and to nonscientific audiences" increased a lot or a little. These gains – addressed mainly through the Science Communication Workshop sessions - were greater than gains reported for other professional research skills, such as "using library database resources" (64%), "condensing literature searches into coherent written introductions," (82%) and "demonstrating new technical skills" (86%).

<u>Note: The Second Cohort</u>

When the pre- and post-workshop survey data for the second cohort – the REU students from the Harvard University were compared to the data from the first cohort, there was substantial similarity although the Harvard cohort showed slightly less enthusiasm overall. Because the data collection from the second cohort did not include the web-based post program survey, administered after the REU program was completed, we have restricted our discussion in this paper to interpreting the data from the first cohort. However, the discussion below includes commentary from Harvard faculty who participated in the stakeholder survey.

<u>Stakeholder interviews</u>

The UMDI evaluator conducted interviews with seven faculty, including the principal investigators from both CHN and the Harvard Nanoscale Science and Engineering Center. He reported: "Feedback overall was very positive. Respondents felt that the workshops were very valuable to students and they wanted to offer similar workshops to their graduate students and even to the faculty in their departments. They felt that the museum staff had been very responsive and actively solicited feedback, and that the workshops were worth disseminating to other museum/university pairings such as those in the NISE network." [7]

Stakeholders also offered several concrete suggestions for improvement for specific aspects of the workshop. As with the student data commenting on specific aspects of the sessions, these will be examined and acted upon by the program team. Our purpose in this paper is to address the overall concept and feasibility of incorporating science communication workshops in to REU programs. Faculty comments concerning the larger issues addressed by the teaching of science communication skills are discussed below.

DISCUSSION

The team of collaborators that designed the new approach to the REU Science Communication Workshops for the summer of 2010, hoped to support faculty goals for the REU program by focusing the sessions on helping students prepare and deliver key deliverables, including a final research presentation suitable for interdisciplinary science and engineering professional audiences. At the same time, the team aimed to help students learn how to seek and describe connections between their daily research activities and the greater societal challenges they may eventually help to address. The team also sought to help students find strategies for sharing their work in science and engineering to broader audiences, including friends and family, as a way of helping them reduce isolation and build supportive social networks. Finally, the team was interested in assessing whether it would be worthwhile to attempt to disseminate the REU Science Communication Workshop model more widely, for integration into other REU programs.

Faculty Perspectives

Faculty stakeholders interviewed by the UDMI evaluator after the program ended showed alignment with these goals, commenting that "students need better oral and written skills for communication with scientists in their own field and in other fields, as well as with the public, and that most scientists receive inadequate training in these areas." Among the necessary communication skills faculty cited were being comfortable speaking in front of audiences, understanding the knowledge and perspective of the audience, avoiding jargon, and presenting in ways that help the audience appreciate and care about the research being presented. Faculty commented that these communication skills would also be helpful during poster sessions at conferences, where students tend to dive too deeply into their material before first addressing the broader goals and context of the research. They thought the workshops helped students see the importance of providing that context.

Good communication skills were seen as particularly important in the field of nanotechnology because "students will often have to address colleagues who aren't familiar with their specific disciplines, because nanotechnology has fewer straightforward applications than other scientific subdisciplines, and because nanotechnology is too small to be seen and therefore requires very clear explanation."

Faculty respondents also commented that "offering science communication training to students also benefits the faculty, department, center, and university," since students communicate better with faculty and with each other. Moreover, when students become more adept at writing and presenting, faculty feel more comfortable sending them to conferences, and they can spend less time revising their papers. Instead, they can provide greater attention to the more technical aspects of scientific training. On a more practical note, faculty stakeholders noted that the workshops helped provide a structure and deadline for the development of the required research presentations, which motivated students to start sooner and produce higher quality work.

Faculty comments also affirmed the value of collaborating with Museum staff in providing the Science Communication Workshops. They noted that REU students tend to listen to outsiders more than to professors, that the museum has credibility with the students, and that it's good for students to see STEM professionals who pursued a career path in education. A change of venue away from the university was also considered to be engaging for students, and gave them "an opportunity, at least briefly, to stop thinking about course grades and to think instead about the big ideas and benefits related to their work". Respondents added that the museum staff are talented, professional presenters who tell good stories, know better than professors how to explain science to the general public, and are 'natural hams' in ways that many professors aren't." [7]

Evaluator's Summary

Before discussing additional feedback on the Science Communication Workshops, we would like to quote the summary from the UMDI evaluation:

In-depth assessment of the science communication workshops offered by the Museum of Science included student surveys, student focus group questions, and phone interviews with key REU program stakeholders. The picture that emerged from these measures was very positive. Students rated the workshops as among the most helpful and enjoyable REU program activities, and their self-reported understanding of how to present their research to both scientific and non-scientific audiences increased from 50% to 94% and 100% respectively. Findings from the phone interviews with key REU program stakeholders were similarly positive. Although they offered several suggestions for next year's workshops, they felt that the 2009 workshops were very valuable for developing the REU students' science communication skills. The faculty wanted to offer similar workshops to graduate students and faculty in their departments, and they encouraged disseminating the workshops to other museum/university pairings in the NISE network. [7]

Additional Student and Faculty Comments

The data show that REU students gained knowledge, skills, and confidence in several aspects of science communication and that they rated the Science Communication Workshops as one of the most helpful aspects of the REU programs. In response to the Web-survey question, "If you were to give an award to the REU program, what would it be for?" several students commended the program for teaching science communication skills, and one participant explained, "[before the REU], I had trouble explaining to other people what I was doing, what I was studying... now I can explain more easily what I'm doing." [7] Another strong affirmation of the Workshops' success in increasing students' self-confidence came in an unsolicited email from a CHN student who had just completed the REU program:

> ... I knew I was going to learn from the [Science Communication Workshops], but I didn't realize how much of an impact it would truly have on me. After I gave my presentation, two of my professors and a peer commented on how drastic of a change there had been not only in the way I presented the work, but also in the way I carried myself. [8].

Consistent with feedback from the faculty stakeholder interviews that even their fellow faculty members could benefit from additional training in science communication, one student commented in his post-program survey, "I wish all the professors and research people would take part in such a workshop. It makes you aware that no matter how great your results are, they don't mean anything if you can't communicate them to others. [9]

Finally, a CHN faculty member sent an unsolicited email suggesting that the Science Communication Workshops substantially improved students' final presentations:

> The results [of the Workshops] really showed in the final presentations, and the students' faculty advisors were pleased... All students had excellent slides, good body language, and no unexplained jargon or filler words... Most also made good eye contact with the audience for the entire presentations... Students clearly explained the background and motivations for their projects, the approaches they used, and their results." [10]

CONCLUSIONS

Data generated by this study show that the REU Science Communication Workshops model described herein has had a beneficial impact on participating REU students and has provided welcome support for the mentoring and educational goals of the REU faculty. Improvements to various program elements can still be made, but the basic concept and integrated approach seems to be working. Key stakeholders agree that the model merits further development and broader dissemination and implementation, through integration into other REU programs, at their own universities and others. To replicate and extend these findings, ongoing assessment of the model's effectiveness with other REU program sites is warranted. This could be carried out in collaboration with local science museum partners, adapting the Workshops to local needs and circumstances, and assisting in the training of local workshop leaders. The abundance of REU programs (see http://www.nsf.gov/crssprgm/reu/reu_search.cfm) and the relative uniformity of REU program structures makes them a promising target for broader dissemination of the REU Science Communication Workshops model to benefit REU students at a critical stage in their education choices and careers.

ACKNOWLEDGMENTS

The authors wish to acknowledge the encouragement and support of the principal investigators of CHN, Ahmed Busnaina and Joey Meade, and the principal investigators of the NSEC at Harvard, Robert Westervelt and Bertrand Halperin. We also wish to thank the NISE Network for additional support and encouragement, and the funders, the National Science Foundation and the Massachusetts Technology Collaborative. Any opinions, conclusions, or recommendations expressed in this material are those of the authors and do not necessarily reflect the views of the NSF or the MTC.

REFERENCES

1. National Science Foundation Program Solicitation 09-598. 2009. *Research Experiences for Undergraduates,* accessible at http://www.nsf.gov/publications/pub_summ.jsp?WT.z_pims_id=5517&ods_key=nsf09598 Also, see NSF REU website for links to a detailed listing of all current programs: http://www.nsf.gov/funding/pgm_summ.jsp?pims_id=5517&from=fund
2. M.P. Hancock, et al. 2008. *Research Experiences for Undergraduates (REU) in the Directorate for Engineering (ENG): 2003-2006, A Draft Report to the National Science Foundation, 2008,* p. ES-4. Accessible at: http://www.google.com/search?source=ig&hl=en&rlz=1G1GGLQ_ENUS350&=&q=reu +engineering+sri+hancock+86%25&aq=f&oq=&aqi=.
3. National Science Foundation. 2007. *Merit Review Broader Impacts Criterion: Representative Activities.* Washington, D.C. Accessible at http://nsf.gov/pubs/policydocs/pappguide/nsf09_29/gpg_index.jsp
4. Alpert, C.L. 2008. *RISE: A Community-Focused Strategy for Public Engagement,* ASTC Dimensions (January/February): 7-8.

5. Westervelt, R. 2008. *Nanoscience and the Public.* ASTC Dimensions (January/February): 8.
6. Tufte, Edward R. (2003), *The Cognitive Style of PowerPoint.* 2003. Cheshire, CT: Graphics Press.
7. UMass Donahue Institute Research & Evaluation Group. 2009. *Center for High-rate Nanomanufacturing Research Experience for Undergraduates: Evaluation of the Summer 2009 Program.* Hadley, Mass.
8. Email communication from REU student to first author.
9. Comment written on an anonymous CHN REU participant survey form.
10. Email communication from CHN faculty member Carol F. Barry.

Mater. Res. Soc. Symp. Proc. Vol. 1233 © 2010 Materials Research Society　　　　1233-PP04-01

Academic/Industrial/NSF Collaborations at the IBM Almaden Research Center- Benefits From Dr. Marni Goldman's Involvement

Charles G. Wade[1], Dolores Miller[1], Kristin Black[2], Curt Frank[2], Joseph Pesek[3]
[1]IBM Almaden Research Center, 650 Harry Road, San Jose, CA 95120
[2] Dept. of Chemical Engineering, Stanford University, Stanford, CA 94305-5025
[3]Dept. of Chemistry, San Jose State University, San Jose, CA 95112-3613

ABSTRACT

College undergraduate and high school teacher internships are a significant factor in materials science education. Traditionally, NSF-supported internships are done in academia. The NSF-supported academic/industrial internship programs involving Stanford University, San Jose University and the IBM Almaden Research Center extend the impact by the inclusion of an industrial research component. Internships through San Jose State University have existed since 1994 under a variety of NSF grants, most recently with NSF-REU support for undergraduate internships. Internships through Stanford university have existed since 1995 through an NSF Materials Research Science and Engineering Center, the "Center for Polymer Interfaces and Macromolecular Interfaces" (CPIMA). In these programs, the interns become members of an existing research group for 10 weeks and have their own project under a mentor. The interns attend a weekly seminar series on industrial research frontiers, a career day, a Graduate Record Examination workshop, a graduate school workshop, and tours of industrial research labs. Every participant presents a poster at an internal technical meeting at IBM at the end of the summer. For the industrial internships at IBM, the research is publishable but closely related to a technical area important to IBM. While the undergraduate and teacher internship programs are the major components of educational outreach of CPIMA, many other projects have been pursued, including public science, programs with local high schools, and science outreach to local community colleges. Dr. Marni Goldman was the Director of Educational Outreach for CPIMA from 2000 until her death in 2007, and she started many of the educational projects and programs. She was especially interested in diversity and initiated an internship program for students who are disabled. The programs will be reviewed and her contributions emphasized.

INTRODUCTION

Two successful NSF-supported academic/industrial research programs involving the IBM Almaden Research Center have shown that the participants (students, postdoctoral scientists, university faculty, teachers and industrial scientists) all derive benefits. In both of these programs, the grants go to the academic partner, and the participants work at IBM and at CPIMA-affiliated industries under legal agreements between the institutions.

One of the programs, an NSF Materials Research Science and Engineering Center (MRSEC), the "Center for Polymer Interfaces and Macromolecular Interfaces" (CPIMA), involves Stanford University, the IBM Almaden Research Center (IBM), the University of California at Berkeley and the University of California at Davis. The CPIMA program which has existed since 1995 has an active group of postdoctoral scientists, graduate students, undergraduate students (summer) and high school teachers (summer) who do research at the academic sites and at IBM as well as some other industrial firms that have some interaction with CPIMA. The second program, with the chemistry department at San Jose State University (SJSU), has had support for 14 of the past 16 years under a variety of NSF initiatives including NSF GOALI, NSF REU programs, and an NSF/DOD joint support REU.

In these programs, each participant becomes a member of individual, existing research groups and does research under the direct guidance of mentors. They attend departmental meetings, seminars and informal discussions. At IBM, the research projects are all publishable but related to a scientific or technical area of interest to IBM. Summer participants at IBM attend a special seminar series on industrial research frontiers, attend department meetings related to research projects and IBM technology, and participate in a variety of other programs sponsored by IBM. Every summer participant presents a technical poster at the end of their stay at a technical forum with typically 100 participants, 8 oral presentations by noted scientists, and 60 posters by interns, graduate students and postdoctoral scientists.

To date, the CPIMA program has placed 140 summer interns and 2 teachers at IBM and the San Jose State University program has placed 260 summer interns and 48 teachers at IBM.

Dr. Marni Goldman, who overcame severe disabilities herself, was the CPIMA Director of Educational Programs and Projects from 2000 until her death in 2007. She was responsible for an array of outreach programs for CPIMA, many of which had an impact on all the NSF-supported summer interns. Her projects included research experiences for teachers, science programs with local high schools and community colleges, enrichment programs with high schools with a high enrollment of disadvantaged students, and a niche program of internships for students with disabilities.

Detailed summaries of the above NSF-supported programs, including policies, recruitment, projects, outreach and benefits through 2001 have been published [1, 2, 3]. The remainder of this presentation will cover outreach activities since 2001 with a focus on Marni's contributions.

OUTREACH PROGRAMS

Goals

The Summer Undergraduate Research Experience (SURE) program has been the flagship of the CPIMA program. The goals of this program are:
- Research experience
- Opportunities for personal growth
- Career Counseling
- Defining point in career as seen from 10 years in the future.

Marni pursued many other programs which became significant to the NSF-supported internships at IBM.

Programs for Students with Disabilities

74

Dr. Goldman started specifically recruiting students with disabilities in 2002 to participate in the CPIMA SURE program, and it evolved with the active participation of the American Association for the Advancement of Science program EntryPoint! which promotes science internships for students with disabilities. Another source has been the National Technical Institute for the Deaf program at the Rochester Institute of Technology. Seventeen students have come to CPIMA in this program. Several have worked at IBM and Stanford, one at the Agilent Corporation and one at UC Davis. A tighter relationship with the Rochester Institute of Technology developed as the program evolved.

Programs for Teachers

In 2002 CPIMA began hosting teachers for 8-week summer research experiences. Dr. Goldman developed a strong teacher professional development program to complement the research component. Teachers spend 1 day each week doing professional development activities including visits to campus research facilities, lectures by a variety of Stanford faculty on current research topics and working with mentor teachers on developing "education transfer plans," based loosely on their summer projects, to use in their classrooms the following year. The other four days of the week, the teachers work on current laboratory research projects under the supervision of sponsoring faculty and graduate student or postdoctoral mentors. In 2005 Dr. Goldman became Associate Director of the Stanford Office of Science Outreach in addition to her duties at CPIMA and expanded the RET program to create a university-wide summer professional development program for science and math teachers. This program continues as the Stanford Summer Research Program for Teachers and is currently supported by an independent NSF RET grant as well as funding from local school districts and from Stanford University funds. Dr. Goldman collaborated with Dr. Jay Dubner (Columbia University) and Dr. Fiona Goodchild (University of California at Santa Barbara) to organize the NSF-supported RETnetwork. She and her collaborators were organizers of the 2002 (Bringing Research Into Science Classrooms), 2003(Assessing, Determining and Measuring The Impacts of the Research Experience) and 2004 RETnetwork conferences. Dr. Goldman also was the webmaster for the organization (http://www.retnetwork.org/).

Community College Programs

Community college students had been included in the SURE program since the 1990s, but Marni developed a special program between Sacramento City College and U.C. Davis. In this program, a Sacramento City College professor and one of his or her students spent the summer at U.C. Davis doing research. Many of these students later matriculated to U.C. Davis. In 2005 she began an outreach program with the Math Engineering Science Achievement (MESA) program at community colleges. MESA is a program which promotes education for economically disadvantaged students at middle school, high school and college levels. She partnered with a faculty member at Skyline Community College to develop a MESA-sponsored evening course in which Skyline faculty members alternated with IBM CPIMA senior investigators and postdoctoral scientists to give talks related to broad scientific research topics: Seeking Knowledge, Generating Knowledge, Communicating Knowledge, and Summer Research Internships. The CPIMA-sponsored sessions included hands-on activities and an informal light dinner to encourage students to get to know the postdoctoral scientists. This program has continued as a stand-alone lecture series that is presented at different local community college MESA centers each year. The graduate students and postdoctoral scientists submit their PowerPoint slides to the CPIMA Education Director for archiving, and the next year's presenters build on the work of previous presenters. Dr. Goldman also organized MESA Day at Stanford in 2006 and 2007 where 100-150 community college MESA students from all over California came to listen to a panel of current and former Stanford students from underrepresented group, hear talks on admissions/financial aid, tour research labs, and tour the campus.

Green Chemistry

Dr. Jim Hedrick (IBM) and Prof. Robert Waymouth (Stanford), in a CPIMA funded collaboration developed interesting research in the field of organic catalysis with special applications to the principal polymer (polyethylene terephthalate or PET) in soft drink bottles. In 2005, a CPIMA SURE intern, Kate Keets, worked with CPIMA graduate student Nahrain Kamber (a former CPIMA SURE intern at IBM) in Prof. Robert Waymouth's lab at Stanford, to develop a plastics recycling lab experiment for undergraduates. This green chemistry experiment demonstrates the recycling of PET materials at ambient conditions.

Collaboration with other NSF and Center Programs

Marni leveraged other NSF grants for collaboration with CPIMA. The National Nanotechnology Infrastructure Network NSF REU program at the Stanford Nanofabrication Facility participated in the CPIMA REU events and the students were housed together. A chemistry REU program at nearby University of Santa Clara, a physics REU at SRI International, and a physics REU at the University of California at Davis were also invited to participate in CPIMA REU events. Many of these participants got special tours of the IBM Almaden Research laboratories. Dr. Goldman also participated in the NSF Research Centers Educator's Network (NRCEN) and was a member of the organizing committee for the 2004 and 2005 meetings. She organized a session on "Diversity in Centers" at the 2004 meeting.

K-12 Programs

In 2002 Dr. Goldman initiated a science fair program at Eastside College Preparatory School, a charter school in East Palo Also with a very high minority population. From 2002 through 2005 CPIMA senior scientists and students from almost all of the CPIMA sites were active in the science fair program. In 2006 Marni began a program of high school internships for Eastside students wherein they participated in a special summer program of 5 weeks at Stanford. This program continues as the RISE (Raising Interest in Science and Engineering) Summer Internship Program, now sponsored by the Stanford

Office of Science Outreach. RISE has expanded to selected other minority-serving high schools and has increased the length of the program to 7 weeks.

Public Science

With support from the Dreyfus Foundation, Marni placed two high school teachers in the summer teachers project at the Exploratorium: the museum of science, art and human perception in San Francisco. The teachers interacted with CPIMA Prof. Vijay Pande at Stanford to develop science education modules on protein folding. Under Pande's direction, the teachers developed an information package which would run on a kiosk computer at the Exploratorium and a mechanical model of protein folding in which magnets on the model simulated hydrogen bonds on the real package. The teachers gave a seminar and demonstration at IBM which also does research in this area.

REFLECTIONS

The summer internships programs have provided for IBM scientists the opportunity to do research not possible without the combined government, university, and industry resources. In addition the CPIMA SURE students have provided an unanticipated benefit: vitality. The flow of new ideas and the poster session are always a highlight of the summer. Marni Goldman's dedication, enthusiasm and educational outreach activities were a force in the program.

ACKNOWLEDGEMENTS

Financial support from the NSF MRSEC Grant 0213618 to Stanford, from the NSF in partnership with the Department of Defense ASSURE program REU grant CHE0552961 to San Jose State University, from a Dreyfus Foundation grant to Stanford, and from IBM is gratefully acknowledged. The authors would also like to thank the participants in the NSF-supported programs for their ideas, enthusiasm, and productivity.

REFERENCES

[1] M. Goldman, C.G. Wade, B.E. Waller and C.W. Frank, *J. Materials Education*, **23**,7-12 (2001)
[2] C. G. Wade, D. C. Miller, M. Goldman, B. Waller, M. Pauly, J. Pesek, and M. Scharberg, *J. Materials Education* 23, 13-27, (2001)
[3] C.G. Wade, D. C. Miller, J. Baglin, M. Goldman, B. Waller, M. Pauly, J. Pesek, M. Scharberg, *J. Materials Education.* 24 , 85-86, (2002).

Mater. Res. Soc. Symp. Proc. Vol. 1233 © 2010 Materials Research Society   1233-PP03-05

Tsihouridis A. Charilaos [1], Polatoglou M. Hariton [2]

[1] School of Special Education, University of Thessaly, Argonafton & Filellinon, 38221, Volos, Greece

[2] Physics Department, Aristotle University of Thessaloniki, 54124 Thessaloniki, Greece

ABSTRACT

In the present work we describe an experimental apparatus of very low cost , aiming at the experimental study of the mechanical properties of solid materials, using the method of indentation. The measurements allow, through a simple analysis, the quantitative determination of the materials' hardness and bulk modulus. A very important class of commonly used materials, namely thermoinsulating materials, is proposed for a case study. The method can furthermore be used to demonstrate the elastic and the plastic behavior of materials. The educational exploitation of the specific apparatus at a technical high school class in Larissa (Greece) and at a university physics students' laboratory practice on the subject of mechanical properties of materials are also described. The responses of the students on the whole process, and specifically on observations related to the pedagogical-educational aspects (active participation, challenge of interest, easiness of measurements, easiness of processing experimental data), as well as related to metrological aspects (uncertainty of measurements), were very positive and are also presented and discussed.

INTRODUCTION

Laboratory activities are defined as learning experiences, in which students interact with models and/or materials in order to observe and understand the natural world and at the same time constitute an excellent medium for the introduction of scientific concepts. It is believed however that meaningful learning can result in only when we give students the opportunity to operate laboratory equipment or materials, thus, participating in the construction of their own knowledge of science and enhancing their conceptual scientific learning. Furthermore, researchers have pointed out that, techniques or methods of laboratory activities should be used by students to investigate phenomena, solve problems or enhance their active participation, positive attitudes and interest on scientific concepts becoming researchers themselves, in contrast to being passive learners [1-6].

In addition, laboratory training is very important, not only for the development of students' practical skills, but also for the development of their cooperative skills which lead to a better conceptual understanding. Above all, it simulates in a way their future workplace, which is very useful, especially for students of technical professions [6-8].

The importance of High School/University students' appropriate laboratory practice has been pointed out by all experts and is considered basic for their education. At the same time, a lot of problems, during the laboratory practice, have been encountered, like: i) possible high cost for the purchase of laboratory equipment ii) difficulty of equipment use iii) lack of appropriate laboratory space iv) complicated apparatuses for measurement v) non understandable operation principles for measurement vi) lack of the possibility to use multiple apparatuses for measurement (due to high cost) vii) no use of the apparatuses for measurement by the students (high cost of possible damage repair) etc.[7]. Consequently, there seems to be a need to conduct experiments with low cost, relevant easiness of use of the apparatuses and have the opportunity to apply metrological methodologies.

The aim of this paper is to set up an apparatus in order to measure the hardness of materials with very low cost, to use it for the measurement of the hardness, as well as the study and the comprehension of the whole process in a school laboratory.

The materials, which were used as a case study, were thermoinsulating materials, which are widely used as building materials in all kinds of constructions. To begin with, the thermoinsulating materials, based on the National System of Accrediting, are defined as the materials, which have thermal resistance greater than 0.1 m^2 K/W and practically have low thermal conductivity. The thermoinsulating materials in this study were given to the authors by various companies engaged in the manufacture and marketing of such materials. All samples studied were of rectangular section 3x12 cm and of 30 cm length.

For a study of the phenomenon usually an expensive and complicated mechanism is required. However, with the possibility of low-cost measurement - in a digital way - of basic magnitudes (length, time), apparatuses can be manufactured, which could allow the measurement of these attributes.

EXPERIMENTAL DETAILS

The hardness of a material depends on its cohesion, as well as its internal structure. Because the results of hardness measurements differ from method to method, every indication of hardness should be accompanied by the trial characteristics. Generally, the different methods of hardness determination give different results because they measure different quantities with different indenters, in a different way. There is not a restricted range for hardness measurement and every trial type has its own determination range of hardness.

The various hardness measurement methods are distinguished into three general categories: a) the dynamic trial methods, b) the static trial methods, and c) the hardness measurement methods with indentation [9-16].

The method used in the present study was the Brinell method, which is most frequently used, especially when materials of certain thickness are tested (and the imprint made on the surface of the material has sizeable dimensions). According to this static method of determination of hardness, a circular hardened steel ball indenter of very hard steel of diameter D and of constant load P, penetrates vertically into the surface of the sample. The ball is pressed smoothly and vertically on the smooth surface of the sample. [9-16].

Description of the Experimental Apparatus and its operation

Our experimental apparatus is based on the method of static indentation, where we have an indention of a steel ball of a certain diameter, under the effect of load P, on the tested sample. The applied load (in N) on the surface produces a circular imprint (in mm^2) which gives the hardness of the material.

Figure 1: Photograph and schematic representation of the simple experimental apparatus of the study of hardness of the thermoinsulating materials.

Firstly, we place the sample on the experimental apparatus and we measure the initial distance of the lever from a fixed point with the help of a micrometer, which is located above the stabilizer of measurement, in an equitable corner. This is accomplished with the stabilization of the micrometer on a rectangular shaped piece of wood, as shown in figure 1.

The weight of the mass of the bob increases, depending on the rule of the lever (figure 1). Thus, a corresponding increase of the applied force on the hardened steel ball indenter can be achieved. Moreover, for the verticality of the indenter on the tested sample, a simple double spirit level is used, which is stabilized on the system of the lever. Its motionless very end can go up and down and can be stabilized on the desirable point with a simple system of screws.

The next step is to place the weights and measure again the distance of the sample from the micrometer. From the difference of the two measurements we calculate the extent of the depth of the steel ball onto the sample. Repeating the above process with different weights we measure the curve of force P(N) vs the depth of the imprint h(mm). Then, a linear approach of the curve follows, as well as a recording of the slope of the resulting straight line.

Measurements

The experimental results that were attained, as well as their elaboration, are all recorded in the following table I. It should be noted here that the measurements were received in different points of the tested sample that abstained between them at least 2 cm. The uncertainties, which could be encountered throughout the whole process, should also be pointed out here (measurement of depth with an electronic micrometer of uncertainty 10μm, measurement of time duration of the weight on the apparatus (30 sec). Other uncertainties could be the measurement of mass with electronic balance of uncertainty of 1 g and its correspondence with weight B in N via its relation with the acceleration of gravity $g = 9,81 m/s^2$, the measurement of the length of the lever in cm and the correspondent point of application of force in the indenter, as well as the uncertainty of the verticality of the indenter on the tested sample).

The average values of the experimental data for one sample and their processing, are aggregated recorded in the table below (table I). The experimental results that were received, as well as their elaboration, for all samples, are graphically presented in figure 2.

We observe that the smallest dipping of the steel ball for the same charge P of our experimental apparatus is observed for sample 7 and thus, it is the "hardest" thermoinsulating sample that was studied.

Table I: Experimental measurements and calculation of hardness BHN (Brinell) for tested sample 1.

a/a	Mass of weight (g)	Total mass (g)	Total Charge (N)	Force P (N)	Dipping (mm) Addition	Dipping (mm) Abstraction	BHN	Diameter of imprint d (mm)	Ratio d/D	P/D
1	230	230	2.26	3.01	0.09		0.56	4.74	0.25	0.006
2	275	505	4.95	6.61	0.38	0.56	0.29	7.81	0.41	0.014
3	230	735	7.21	9.61	0.87	1.13	0.19	9.38	0.49	0.020
4	230	965	9.47	12.62	1.03	0.88	0.21	10.49	0.55	0.026
5	305	1270	12.46	16.61	1.46	1.35	0.19	11.63	0.61	0.035
6	320	1590	15.60	20.80	2.06	1.27	0.17	12.99	0.68	0.043
7	330	1920	18.84	25.11	2.29	1.56	0.18	13.54	0.71	0.052
8	375	2295	22.51	30.02	2.48	1.81	0.20	14.56	0.77	0.062
9	395	2690	26.39	35.19	3.22	2.04	0.18	15.18	0.80	0.073
10	400	3090	30.31	40.42	3.72	2.43	0.18	15.95	0.84	0.084
11	420	3510	34.43	45.91	4.21	2.64	0.18	16.71	0.88	0.095

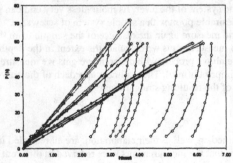

Sample 7: 20.124x-0.7078
Sample 6: 19.142x-0.1174
Sample 5: 14.741x+0.8813
Sample 4: 13.723x+1.1216
Sample 3: 9.7509x+2.2856
Sample 2: 9.3663x+1.8978
Sample 1: 9.8761x-0.0009

Figure 2: Diagram of load vs dipping during the loading and unloading for all tested samples

Instructive intervention-evaluation

In this stage an instructive intervention was designed and carried out, with the use of the apparatus, in a group of university students and two groups of high school students, in order to introduce them to the concepts of metrology and mechanical attributes. The whole process was evaluated, on the one hand from an instructive and pedagogic point of view (active participation, challenge of interest, easiness of measurements, easiness of processing of experimental data, achievement of objectives) and on the other hand from a scientific point of view (uncertainty of measurements).

Intervention phase I

It was conducted with a group of students of the Physics Department of the Aristotle University of Thessaloniki including: a) the demonstration and analysis of the experimental apparatus and its way of operation, b) the carrying out of measurements c) an open discussion and recording of their opinion, with regard to the whole process, as well as the evaluation of the apparatus.

It is worth mentioning that the intervention to the group of students lasted for an hour, due to their fast familiarization with the apparatus and the taking and processing of the measurements.

Intervention phase II

It was conducted with two groups of four students each, of a Technical Vocational Secondary School, (last class), which were given different thermoinsulating materials to study and evaluate. The first group attended the Mechanics specialty course having the knowledge of the mechanical attributes of materials (they had been taught relative subjects during the previous school years) and the second one attended the Electronics specialty course with no knowledge of the mechanical attributes of the materials, other than their experiences.

The intervention was based on a lesson plan, which lasted for three instructive hours.

1st instructive hour: a) presentation of the relevant theory b) demonstration and explanation of the operation of the apparatus c) explanation of measurement taking and processing

2nd instructive hour: a) The taking of measurements of each group, where each student's role was different b) processing of measurements with the help of the Excel program and drawing of the experimental curves for each corresponding material

3rd instructive hour: a) presentation of the experimental curves of each group's material b) comparative study of the curves c) semi-structured interview with each student and recording of their opinion with regard to the whole process and its final evaluation by them

DISCUSSION AND CONCLUSIONS

The analysis of the university and high school students' comments showed, that there was a direct familiarization with the whole apparatus, due to the easiness of use (addition of weights and measurement of depth with the electronic micrometer), easiness of measurements processing (graphic representation of the P(N) curve vs h(mm) and its linear approach with the help of the Excel program.

Furthermore, the groups of the university and high school students commented positively on the whole process, due to the repetition of the measurements. With the repetition of the whole process of the measurements there was a direct correlation of the depth of the imprint, with the corresponding force of the charge and the hardness of the material (small dipping depth in harder materials), as well as an easier comprehension of the concept of material hardness. Finally, we consider that from a pedagogical point of view, the apparatus achieved its purpose, as there was an active participation of all group members and since, for the accomplishment of measurements, the harmonious collaboration of all participants was needed, due to the distinguishable role of each student, as mentioned above.

As well, the correlation of the slope of the linear approach of the curve P(N) vs h(mm) with the hardness of the measured material, was considered by the students simple and comprehensible. It should be particularly stressed, that there was no differentiation, as concerns the comprehension of the concept of hardness, in the two different groups of high school students (of the Mechanics and Electronics Specialty) and the experimental results were absolutely comparable among them. The students carried out the experimental measurements and their processing with easiness and finally compared two different materials, as regards their hardness. Consequently, one could say, that the initial objective of the apparatus was achieved, at least at this stage.

The second objective of the whole apparatus was to introduce the students to the metrology of the mechanical attributes of materials. For this objective, explanations were initially given for the concept of uncertainty and the obvious uncertainties, which could be confronted during the process, were pointed out. The university students mainly showed that they comprehended the concept of uncertainty and its role in measurements, while there was some difficulty for the students of the Secondary School.

Conclusively, it should be noted that the initial objective was to set up a device- an experimental apparatus - with particularly low cost, that presents easiness of use, helping mainly the students to comprehend the mechanical attributes of materials (hardness, elasticity), as well as to introduce them to the metrology of the above mechanical attributes. The results of the relative instructive interventions, as well as the participants' opinions, which were expressed via the interviews, showed that the university and high school students reacted positively to the use of the experimental apparatus, having the opportunity to experiment with the measurement of the hardness of various materials, and have a better understanding of the related physical phenomenon.

REFERENCES

1. A. Hofstein and V. N. Lunetta, *Sc. Educ.* 88(1), 28-54, (2003).
2. W. M. Roth, *Res. in Sc. Teach.* 31, 197-223, (1994).
3. K. G. Tobin, *Sch. Sc. and Math.* 90, 403-418, (1990).
4. J. R. Baird, in *The Student Laboratory And The Science Curriculum*, edited by E. Hegarty-Hazel (London:Routledge, 1990), pp. 183-200.
5. D. Hodson, *Stud. in Sc. Educ.* 22, 85-142, (1993).
6. R. Lazarowitz and P. Tamir, in *Handbook Of Research On Science Teaching And Learning*, edited by D. L. Gabel (New York:Macmillan, 1994), pp. 94-130.
7. Ch.Tsihouridis , D.Vavougios, and D.S. Ioannidis, *Proceedings of 12th International Conference on Interactive Computer Aided Learning / ICL2009*, (Auer M. (Eds.), Villach, Austria, 2009, Kassel University Press ISBN 978-3-89958-481-3), pp. 795-811.
8. A. J. Cox and W. F. Junkin, *Phys. Educ.* 37(1), 37-44, (2002)
9. JM Gere and S. P. Timoshenko, *Mechanics of materials*, 4th ed. (Stanley Thornes Publishers, 1999).
10. D.Tabor, *The Hardness of Metals* (Oxford University Press, 1951).
11. M. M. Eisenstadt, *Introduction to Mechanical Properties of Materials*, (MacMillan, 1971).
12. Z.H.Stachurski, *Mater. For.* 30, 118-124, (2006).
13. W. G. Moffat, T. H. Courtney and J. Wulff, *The Structure and Properties of Materials*, (Vol. 3, John Wiley & Sons, 1967).
14. H. Chandler, *Hardness Testing*, 2nd ed. (ASM International, 1999).
15. R. Hibbeler, *Mechanics of Materials*, (Prentice Hall, New Jersey, 1997).
16. A. Higdon, S. Ohlsen, W. Stiles, J. Weese, and W. Riley, *Mechanics of Materials*, (John Wiley and Sons, 1978).

Mater. Res. Soc. Symp. Proc. Vol. 1233 © 2010 Materials Research Society   1233-PP09-06

Dolores C. Miller[1], Frances A. Houle[2], Janet Stemwedel[3], Joseph Pesek[4] and Charles Wade[1]

[1] IBM Almaden Research Center, 650 Harry Road, San Jose CA 95120, U.S.A.
[2] formerly IBM Almaden Research Center, 650 Harry Road, San Jose CA 95120, U.S.A.
[3] Dept. of Philosophy, San Jose State University, 1 Washington Square, San Jose CA 95112, U.S.A.
[4] Dept. of Chemistry, San Jose State University, 1 Washington Square, San Jose CA 95112, U.S.A.

ABSTRACT

As part of the NSF supported SUMMIT (San Jose State University/IBM Almaden Research Center collaboration) REU and CPIMA (Stanford University/Almaden) SURE undergraduate internship programs we have developed a workshop that emphasizes the scientific community's dependence on ethical behavior for its success and advancement at this nascent stage of a scientific career. We have successfully introduced open-ended role playing based on recent real life ethics cases as reported in the scientific news to a foundation presentation based on the APS Ethics and Professional Conduct Guidelines for Physicists. The goal is to foster discussion of complex cases and their effects not only on the protagonists but on the scientific community.

INTRODUCTION

Ethics in science research and education continues to be a subject of much concern. Studies, task forces and committees from various organizations have tried to address best practices in defining and teaching ethics to students at both the undergraduate and graduate levels [1]. The National Science Foundation deemed ethics training important enough that separate supplemental funding was made available for developing ethics modules for participants of the Research Experience for Undergraduates (REU) programs in their 2004 call for proposals [2].

Since participating in an REU is often a student's first experience as an active member of the research community, introducing them to aspects of scientific research complementary to their technical work has long been an integral part of the summer programs that bring undergraduate interns to the IBM Almaden Research Center. The mandatory weekly summer undergraduate intern seminar series has included practice in extemporaneous technical presentation and a technical poster presentation workshop, for example. Given the importance of the subject, an ethics module was developed and presented as part of the weekly undergraduate seminar series. Our overarching goal was to foster thought about and discussion of the dependence of the scientific community on ethical behavior. Our program consisted of a foundation lecture, a role playing exercise based on real ethics cases and a subsequent report back and general discussion of the role play.

ETHICS MODULE DESCRIPTION

Participants at Almaden were from the NSF supported Center for Polymer Interfaces and Macromolecular Assemblies (CPIMA) SURE (in collaboration with Stanford University) and the Analysis, Fabrication and Engineering of Surfaces and Materials in Information Technology SUMMIT REU (in collaboration with San Jose State University) programs as well as two IBM supported programs for women and underrepresented minorities. All were science majors in chemistry, chemical engineering, physics and related sciences or secondary school science teachers. About half of the undergraduates had done some scientific research, either at their home institution or at a previous REU. In 2009 a session was also held at Stanford as part of the Stanford CPIMA SURE seminar series, in which participants from other undergraduate/teacher science research programs were invited to participate. Few interns reported having had previous research ethics training; those that did had it in a bio/medical context.

Given the short time available (one sixty minute and one ninety minute session in the final module) and the academically diverse backgrounds of the interns the module was designed to be an introduction to the major ethical issues pertinent to research in the physical sciences rather than an exhaustive survey. A foundation presentation was given that was based on the APS Ethics and Professional Conduct Guidelines for Physicists [3]. One of the co-authors (Houle) had served on a 2003 to 2004 APS task force that was created in response to two high profile physics ethics cases. The interns were also given the formal definition of research misconduct as described in the National Science and Technology Council's Federal Policy on Research Misconduct [4]. An overview of expectations beyond avoiding formal misconduct- truthful, careful handling and reporting of data; responsible, respectful interactions with colleagues and subordinates; confidentiality in refereeing and reviews; honesty in publication; proper recognition of research contributions of colleagues; complete citation of prior work; no multiple publications of same work; disclosure and avoidance of conflicts of interest, responsible use of resources - underscored the responsibilities of belonging to the scientific community. Ethical best practices for mentors (teaching standards for high quality technical work, discussing ethical behavior, developing a culture of community and wise use of resources, promoting timely professional advancement and acknowledging contributions) were described, giving interns guidelines for what to look for in an excellent research advisor or other supervisor. The importance of the research record (and the changes generated by the emergence of electronic data acquisition and storage) was discussed, as were the rights and responsibilities of co-authorship.

Finally, the interns were given guidelines on what to do if something doesn't seem right, including talking to the people involved to clear up any misunderstandings and discussing the situation with the institution's ombudsman or professional ethics officer. They were informed that all scientific institutions are required to have someone on staff with the responsibility of handling ethics inquiries and advised to let the institution go through its process. Throughout the presentation actual cases and events were used as examples. Handouts of the presentation had been provided at the beginning of the session and there was an opportunity for questions at the end.

For the first two years simple fictional academic case studies were used as the springboard to discussion. Intern participation was minimal and responses had to be elicited. We found this somewhat surprising because the group had been reasonably voluble in discussions of a wide range of technical material with which they had had little familiarity. It was subsequently decided to use role playing of ethics situations to try to engage the interns in active participation. There are various suggestions for role playing case studies on the web [5] and in the literature

[6]. We chose to use actual recent ethics cases involving currently active investigators as reported in the scientific news [7, 8] to bring the issues alive and show their relevance to the real scientific world. The selection criteria for the articles were that the research in question be in areas in the general sphere of the interns' academic backgrounds and/or summer technical work; that many viewpoints be expressed, particularly one that might be considered representative of the scientific community as an entity; and, perhaps most importantly, the outcome be ambiguous or controversial- having no one "right" answer.

The final structure of the module in years three and four was that in week one the interns were given the articles to read, with no directions other than that they were free to discuss them amongst themselves and their mentors. This was done to avoid having them read "to the test"- to let them think about the articles and form their own ideas based on their background and understanding. The ethics in research presentation described above was given in week two. The role play process and discussion questions/guides (Figure 1) were introduced at this point.

Actions?

Motivation?

Result(s) for you? Impact on others?

What remains unresolved?

What would /could you do differently?
With what ramifications ?

Figure 1. Role-play guidelines (shown above), a summary of the characters' position, relationship to the case protagonist, and description and information about the character (according to the author of the news article) were placed on the character place cards as reminders and discussion aids.

From four to six parties with significant coverage in the news article had been chosen as characters in the role play for each article- the party under investigation and the initiator of the investigation, of course, plus other interested/affected parties (colleagues, students and ethics investigators, for example). These characters were randomly assigned to the participants who could trade roles if they desired. All of the Almaden interns were required to participate in the exercise, as they were for the other practice activities in the seminar series.

The actual role play was done in the third week. The two cases used involved a researcher accused of falsifying data and subsequently accusing his graduate student of fabrication when a second research group could not reproduce the results [7], and another who purportedly mimicked the experiment of a researcher in his field whose lecture he had attended and rushed to get his results into the literature first [8]. Each group of interns representing the parties from a single case independently discussed, starting in their character's voices, the ethical issues of their case using the formal presentation as a foundation and the guidelines (Figure 1) as the starting point. There were four groups total, two for each case at each session. A facilitator from the program staff sat in with each group throughout the role-play to answer questions, keep the discussion moving, play devil's advocate as necessary and make sure all participants had a voice. Because the goal was to have the participants practice grappling with ethics (rather than to come to some predetermined "correct" answer on a specific issue) we chose to let each discussion follow its own dynamic.

Each role-play group prepared a summary outlining the path and results of their discussion and chose a presenter to report back to everyone. These presentations were open to comment by the other groups. The facilitators underscored the common themes across the various presentations, made clarifications and gave more food for thought. The wrap-up of the session was a discussion of the seeds of ethical misbehavior, as described in an editorial by D. Goodstein in Physics World [9]. A source list for further reading on research ethics was provided. Finally the interns were encouraged to talk about the session and research ethics in general with their technical mentors.

DISCUSSION

The role-play led to very active, diverse discussions. Some of the facilitators tried to keep the participants "in character" throughout the discussion, while others used the role-play only to introduce the situation as presented in the article. This was not considered a problem since the goal was to get a discussion going rather than to strictly follow a procedure or solve a specific problem. Having a facilitator for each group was labor intensive but crucial to maintaining a small comfortable discussion environment.

The summary presentations of the role-plays and discussions demonstrated the different emphases of the discussions. Some examples are listed in Table I. There are some common points in the reports for each article, and it seems clear that the participants were drawing on the foundation ethics lecture to form the framework of their discussions. Almost every report back included some consideration of the scientific community as a whole, either in terms of ethical responsibility or of being affected by unethical behavior. There were also some interesting lines of discussion that seemed to be outside of the information presented, such as who pays for an ethics investigation. One group even questioned and discussed the evenhandedness of the news article coverage of the ethics case.

One drawback to using a real case with an accused ethics violator was that a couple of participants playing that role identified them as the villain (behaving unethically because they were inherently evil) and therefore had difficulty discussing the case from that person's viewpoint. It was important that the participants understand the assumptions behind most unethical behavior (pressure to be first, "knowing" one is right before confirmation from data, for example [9]) in order to recognize the situations and temptations that lead to it, so some effort was spent in drawing that character's presumed motives out in those cases.

CONCLUSION

There was much more engagement in the ethics module after the introduction of the role-play. We received informal positive feedback on the usefulness of the ethics module from the interns. Mentors confirmed that the interns took the initiative to bring up the subject of ethics within their technical groups. Several interns were interested enough to follow up on the current status of the researchers investigated in the role-play articles.

Case [ref]	Discussion Points
1[7]	Communications between graduate student and advisor Chain of responsibility Responsibilities of an author Importance of verification; Importance of collaboration Bias in presentation [of article] Ethical responsibilities and the pitfalls of the feudal organization [research group]
1[7]	Unresolved: blame > P.I. or grad student Responsibility [for] the data Veracity vs. funding/publication/prestige Check your DATA, question things Focus on the situation, not the people Responsibility after publication
2[8]	Severity of ethical violations, punishment Stealing/failure to cite vs. fabrication Who is responsible for catching violations- journal reviewers, community Who teaches ethics- role of mentor, teaching vs. observation Someone's career is ruined, either [researcher in case 2] or [complainant in case 2] Everyone is responsible for policing self
2[8]	Punishment should have been more severe Need more protocols/committees to deal with these situations Have undergrads/grad go through practical ethics classes before participating in research Investigations can be a waste of time so some scientists may get away with fraud (there needs to be a strong incentive to investigate)
2[8]	Should be able to be confident when presenting work to peers Hard to fire tenured professor for borderline [ethics] cases [Researcher in case 2] feels he is not doing anything wrong; does not seem to care Common motivation is FAME Reputation should be soiled; still gets money, students, publications Who pays for investigation? You do NOT get to the top alone

ACKNOWLEDGMENTS

Support from the Department of Defense ASSURE program in partnership with the National Science Foundation (Grant CHE0552961) for the SUMMIT REU program and from NSF (grants DMR0213618 and DMR0243886) for the CPIMA SURE program is acknowledged.

REFERENCES

1. Rachelle Hollander, editor; Carol R. Arenberg, co-editor, Ethics Education and Scientific and Engineering Research: What's Been Learned? What Should Be Done? Summary of a Workshop, (The National Academies Press, Washington D.C. 2009), for example.

2. Research Experiences for Undergraduates Supplements and Sites NSF Program Solicitation 04-584 (2004).
3. APS Guidelines for Professional Conduct, (Adopted on November 10, 2002) http://www.aps.org/policy/statements/04_1.cfm
4. National Science and Technology Council, Federal Policy on Research Misconduct, http://www.ostp.gov/cs/federal_policy_on_research_misconduct .
5. http://beta.onlineethics.org/Resources/Cases.aspx and cases cited therein.
6. Committee on Science, Engineering, and Public Policy, National Academy of Sciences, National Academy of Engineering, and Institute of Medicine, On Being a Scientist: a Guide to Responsible Conduct in Research, (The National Academies Press, Washington D.C. 2009)
7. E. C. Hayden, Nature News **453**, 275-278 (2008).
8. W. G. Schulz, C&E News **85** (12) 35-38 (2007).
9. D. Goodstein, Physics World **15** Nov. 2002, 17.

Mater. Res. Soc. Symp. Proc. Vol. 1233 © 2010 Materials Research Society   1233-PP03-04

Discovery Learning Tools in Materials Science: Concept Visualization With Dynamic and Interactive Spreadsheets

Scott A. Sinex[1] and Joshua B. Halpern[2]

[1]Department of Physical Sciences and Engineering, Prince George's Community College, Largo, MD 20774 and [2]Department of Chemistry, Howard University, Washington, DC 20059

Abstract

Many materials science concepts can be developed into animated, interactive spreadsheets to create engaging discovery learning tools. These Excel spreadsheets do not require programming expertise. Learning how to create and use these didactically useful spreadsheets is simple and new examples can be quickly created by instructors.

Introduction

Are the bond lengths in a metal or an ionic compound *constant*? How are bond lengths measured? These are examples of two rather simple questions that students can discover answers for themselves via dynamic and interactive spreadsheets or Excelets (Java-less applet-like). The MatSci Excelets collection [1] of over thirty dynamic and interactive spreadsheets offers instructors in introductory materials science or other courses such as general chemistry or physics, a way to create an engaging pedagogy in the classroom, whether for lecture, laboratory, or out-of-class assignments. The spreadsheets are all computationally-based (using formulae, not programming) and easily modified by instructors just being introduced to the technology and even by students interested in exploratory learning. This collection complements the Spreadsheet Applications for Materials Science, SAMS, by Meier [2]. This paper describes the pedagogical use of some of the MatSci Excelets in the classroom and student feedback.

The Solid State – A Discovery-based Lecture-Discussion

Here are a series of examples that could be presented in lectures, but allow students to discover the answers guided by the instructor posing questions. The first example, shown in Figure 1,

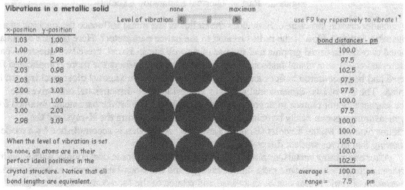

Figure 1. Vibrations in a metallic solid

deals with the question of whether bond lengths in a metal are constant. It starts with all the bond lengths at a distance of 100 pm. Instructors can create vibration at different levels using the scroll bar and hitting the F9 key multiple times, animating the atoms. By increasing the level of vibration, mimicking a temperature increase, the variation in bond lengths increases. This is a simple but nevertheless, a very powerful visualization of a microscopic (atomic) process for novice learners.

Thermal expansion is another important concept in materials. This seemingly simple macroscopic concept can be confusing for students, but some of these issues can be addressed using Excelets. For example, as shown in Figure 2, what is the difference between an isotropic and anisotropic material? What happens if the temperature increases? Here an instructor increases the temperature to cause thermal expansion and students view how the material behaves. They can then deduce the difference between isotropic and anisotropic expansion, as opposed to an instructor just flat out telling them.

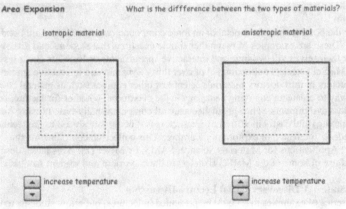

Figure 2. Part of Thermal Expansion Excelet

Now consider crystal structures. How does the radius of the metal influence the packing in a simple cubic lattice? How is the radius related to the lattice parameter? Here (Figure 3) via an animated two-dimensional graphic and the use of a scroll bar in Excel, these two questions can be addressed easily in a visual fashion. Then in the same spreadsheet the more complicated face-centered and body-centered lattices and on another spreadsheet hexagonal close packing can be explored. The use of two-dimensional graphics alongside three-dimensional ones gives the novice learner a better chance to develop an understanding. The lattice parameter obtained from x-ray measurements can easily be related and can be explored using the X-rays and the Crystalline State of Matter Excelet (Bragg's Law) too. All of this is accomplished by a mode of questioning that drives student discovery.

After examining metals, exploring the alkali halides, all ionic compounds with sodium chloride structures, allows ionic radii, lattice energy, and the Born-Haber cycle to be covered. The variation of ionic radius can be examined and related to the face-centered cubic crystal

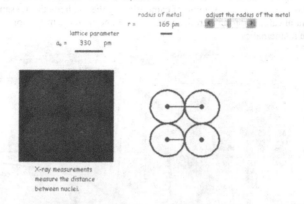

Figure 3. Simple Cubic Lattice

structure. Now the trends of the lattice parameter (a typical crystallographic measurement) as a function of anion size can be explored graphically (Figure 4) using the spreadsheet to handle a small database and allowing student control (check boxes on the right of the screenshot) of the data on the graph. Students can discover how the ionic radii influence the trends as both cation and anion sizes vary. More of the database aspects can be found in Exploring the Periodic Table: Metal-Nonmetal Comparison.

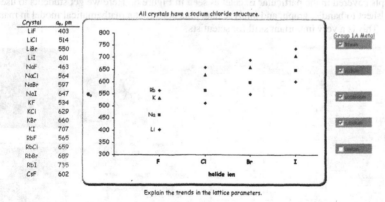

Figure 4. Exploring the Lattice Parameters

The Bonding in Sodium Chloride Crystals Excelet demonstrates the use of graphing the calculation of the attractive and repulsive energies for sodium chloride (Figure 5). Using a tracer line to find the minimum energy on the net energy curve, which is the sum of the attractive and repulsive energies, students can discover the radius (sum of cation and anion) that minimizes energy. This is an excellent example of camouflaging the mathematics (essentially the Lennard-Jones potential calculation) to introduce the concept and then the instructor can decide how far to bring in the mathematics.

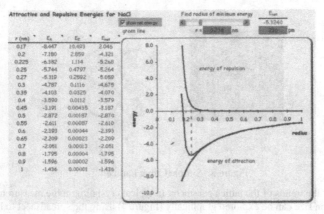

Figure 5. Bonding in sodium chloride structure

Out-of-class Projects

Many of the MatSci Excelets contain assessment questions using data that expand on concepts covered in the particular Excelet as seen in Figure 6. Here we get students to use the spreadsheet to handle, graph, and analyze data plus construct a mathematical model in many cases. This is a very important skill for scientists.

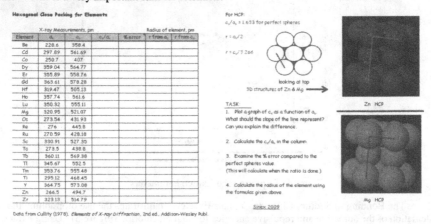

Figure 6. Assessment Questions

At present, there are 32 spreadsheets in the collection. The ones that involve crystal structures also have associated web pages with Chime structures (interactive 3D structures). Many of the topics such as Brinell hardness, the Avrami equation, and three-point load test for fracturing ceramics are covered in typical sophomore level textbooks for materials science such as Callister [3]. All of the MatSci Excelets are freely available for download at http://academic.pgcc.edu/~ssinex/excelets/matsci_excelets.htm.

Student Feedback

The interactive nature of Excelets was favored because it allowed students to play and experiment (18%), discover on their own (18%), think (7%) or all three of these (56%). In general, the visual aspects were appreciated and students felt it enhanced their understanding of concepts. Table 1 below summarizes four questions from a survey of 28 students in general chemistry in spring 2009.

Table 1. Student Feedback

Which do you prefer:		
static graphs in textbook	7%	(2)
dynamic graphs in Excel	89%	(25)
both	4%	(1)
Using Excelets does not require that you be familiar with Excel.		
most definitely	25%	(7)
I think so	36%	(10)
just barely	32%	(9)
not at all	7%	(2)
don't know		
Excelets offer a more visual experience with graphs instead of using just the mathematical equations.		
most definitely	75%	(21)
I think so	14%	(4)
just barely	14%	(4)
not at all		
don't know		
Excelets make it easier to grasp or learn a concept.		
most definitely	43%	(12)
I think so	54%	(15)
just barely	4%	(1)
not at all		
don't know		

Students were asked to rank how Excelets were used in the class. The results are listed in order from most (1 of 5) to least (5 of 5) favorable: group activity done in class (1.8) > lecture using Excelet (1.9) > group activity done out of class (3.2) > individual project (3.3) > lecture with no Excelet (4.0). This reflects how the students wanted the instructor to use the Excelets.

Constructing Interactive Excel Spreadsheets or Excelets

The wherewithal for producing Excelets can be found at the Developer's Guide to Excelets [4], which includes a tutorial, illustrated instructions, and many more examples. We strongly encourage others to take the interactive features tour to see what can be done in Excel. The forms toolbar provides a variety of features (spinners, scroll bars, checkboxes, etc.) that are easy to use, and when combined with logical functions, lookup tables, conditional formatting, and a number of simple tricks provides a wealth of interactivity and dynamic display. The use of the forms toolbar allows Excelets to function on both PC and Mac platforms. All of this is done using computations (a.k.a. - formulas and available functions) in the cells. The use of comment boxes adds explanation, hints, and answers for students as well. One should always look under the graphs, as many of the tricks, such as turning lines on graphs on and off or tracer points are explained there. The support materials for mathematical modeling are also given. This includes modeling and simulations with cookies, introduction to linear regression, goodness of fit for linear models, interpolation and extrapolation, and non-linear models as well.

Many of the Excelets will be completely functional in Open Office Calc, which is open source (free) software available at http://www.openoffice.org. Calc functions very much like Excel. One may have to clean up the files for cosmetic reasons. Check boxes will become colorless and the words true and false which are under them need to be hidden (turn the font white). If option buttons are in the Excelet, they have a slightly different syntax in Calc and the formula must be amended in Calc for the spreadsheet to function.

Some Final Thoughts

Excelets provide a "click-and-think" mode of learning for students in a large variety of topics in introductory materials science and general chemistry using an off-the-shelf software package. Even in a lecture environment, the "chalk-and-talk" mode can be minimized by using Excelets. Classical lectures can evolve into a more discussion-based format with students discovering concepts through predict-test-analyze (the "what if" scenarios) using technology as well. Student engagement via out-of-class projects or follow-up activities in the laboratory is increased considerably. Excelets also provide another great way to add interactivity to instruction in online classes.

Excelets address the visual (various types of graphs, use of conditional formatting) and kinesthetic (interactive features) learning styles. With the mode of questioning, they drive students toward deeper learning using an engaging pedagogy.

This work is supported by the Howard/Hopkins/PGCC Partnership for Research and Education in Materials (PREM), funded by NSF Grant No. DMR-0611595.

References

1. The MatSci Excelets collection is freely available for download at http://academic.pgcc.edu/~ssinex/excelets/matsci_excelets.htm (accessed January 2010).

2. Meier, M.L. (2003) Spreadsheet Applications for Materials Science, Proceedings of the 2003 American Society for Engineering education Annual Conference. Collection of spreadsheets available at http://www.kstreetstudio.com/sams/index.htm (accessed January 2010).

3. Callister, W.D. (2006) **Materials Science and Engineering: An Introduction**, 7th Edition, Wiley.

4. Sinex, S.A. (2009) Developer's Guide to Excelets: Dynamic and Interactive Visualization with "Javaless" Applets or Interactive Excel Spreadsheets, http://academic.pgcc.edu/~ssinex/excelets (accessed January 2010).

Mater. Res. Soc. Symp. Proc. Vol. 1233 © 2010 Materials Research Society   1233-PP04-32

Authentic Science Research and the Utilization of Nanoscience in the Non-Traditional Classroom Setting

Deborah A. Day[1], Zizi Yu[1], Zelun Wang[1], Jennifer Dalecki[1], Arian Jadbabaie[1], Emily Z. Feng[1],

Thomas J. Mattessich[1], Christine Caragianis-Broadbridge[2], Mark A. Reed[3], Ryan Munden[3]

[1] Amity Senior High School, Woodbridge, CT;
[2] Department of Physics, Southern Connecticut State University, New Haven, CT;
[3] Department of Applied Physics, Yale University, New Haven, CT

ABSTRACT

Applications of nanoscience in the non-traditional classroom have successfully exposed students to various methods of research with applications to micro- and nano-electronics. Activities obtained from the NanoSense website associated with current global energy and water concerns are solid examples[1]. In this regard, all 36 students in the 2008-2009 Science Research Program (SRP) prepared and delivered individual and group lesson plans in addition to their authentic, year-long research projects. Two out of 36 students selected nanoscience based projects in preparation for science fair competition in 2009. Additionally, preliminary research was conducted while participating in the Center for Research on Interface Structures and Phenomena (CRISP) Research Experience for Teachers (RET) Program in summer 2008 which supported the idea of developing a photolithography kit. This kit is intended to introduce high school students to the fundamentals of photolithography. In this paper, the design, implementation and feasibility of this kit in the high school classroom is described as well as details involving individual and group nanoscience based projects. Supporting educational models include self-regulated learning (SRL) concepts; situated cognition; social constructivism; Renzulli's (1977) enrichment triad and Types I – III inquiry enrichment activities[2,3].

INTRODUCTION

The philosophy behind the Authentic Science Research Program (SRP), a non-traditional high school honors elective course originally created by Dr. Robert Pavlica, is to provide students with an understanding of scientific research methodology by way of extensive bibliographic and laboratory experimentation. In this setting, students consult with doctoral level scholars in planning long-term focused research to investigate an authentic research question. In doing so, students demonstrate initiative, perseverance, creativity, originality, ingenuity, intuitiveness, motivation and self-regulation. The program affords students the opportunity to participate in the community of scientific research and scholarship as part of their high school experience by first selecting a topic for investigation, reviewing current scientific literature, securing a mentor, engaging in an original piece of research, gathering data, conducting statistical analyses on these data, drawing conclusions and communicating their findings in oral, poster and written form. Involvement in cutting-edge research is the essence of the program and is accomplished not only through individual research but group-oriented projects as well[4].

THEORY

Self Regulated Learning (SRL) is a process whereby the learner autonomously takes control of their learning process[2]. Specific, concrete techniques identified by the biofunctional self-regulation theory (Iran-Nejad, 1990) assist SRL[5]. Reading for comprehension and note-taking techniques require students to utilize higher order cognitive strategies in order to identify suitable research questions, monitor content understanding, and summarize information. Coupled with one-on-one SRL facilitated meetings with an instructor and regular class attendance, students who are successful with this methodology gain not only the skills necessary for academic achievement, but have been shown to possess the will to do so[2].

Literature suggests that SRL is the basis by which academic achievement, as defined by track placement and overall standardized test scores, is promoted and that the self-regulated process, according to Iran-Nejad allows for an improvement in long-term memory. It is suggested that this long-term memory is accomplished through active, self-regulation; selective attention; self-questioning; prediction and procedural metacognition (the ability to think about thinking). Both traditional and contemporary understanding of SRL is empirically identified under social cognitive theoretical framework and can be directly associated with student's cognitive self-regulation within the educational setting (Puustinen and Pulkkinen (2001))[6].

A learning environment that promotes SRL at the high school level has advantages such as improved degree completion rates on the college level. In addition to this self-directive process, the implementation of situated cognition, student learning as part of a community, provides additional educational benefits. In this model, a student moves from the role of an observing newcomer to the role of an active participant within the (scientific) community. To achieve this open inquiry experience – one in which the student identifies a quality, researchable question then implements the scientific method in order to suggest a credible solution – the student must design and conduct experiments to test his or her hypotheses related to the authentic research question. An active, student-centered, community based environment provides the social component necessary whereby learning takes place in a co-constitutive manner. Here, students not only learn from instructors and/or experts, but instructors and experts learn from students. Thus, knowledge and skills are acquired within a real life context, promoting a community of learners and "culture of practice"[7].

The research of Roth and Roychoudhury (1993) concluded that students were better able to define concepts, events and actions to design their experiments and communicate their results when placed in a situated context[8]. In this regard, the social constructivism movement has challenged more traditional models of education and places knowledge within the process of social interchange. This, in turn, may prove valuable for insight regarding scientific inquiry[9].

Renzulli's (1977) enrichment triad model was originally developed for gifted education in the primary schools[3]. Follow up research (Renzulli & Reis, 1986) deemed this model transferable to secondary education. Three types of enrichment activities, all interrelated, were identified as follows: Type I activities are categorized as exploratory in nature with a focus on exposing students to new possible areas of interest. To accomplish this, printed materials, guest speaker visitations, field trips and pointed Internet activities are all modes of exploration. Type II activities best represent "how to" accomplish a particular task -- with a focus on technical and cognitive skill building. Activities include creative and critical problem solving, decision making, communication skills, affective processing and general research techniques. Type III activities encourage autonomy and ownership of learning by way of student role as information producer rather than consumer. Identifying and solving problems are critical components in this type of enrichment. Communication and presentation skills are at the highest possible level in the process of emulating professionals in a chosen field. Curriculum support, Type I and II activities, and environmental and/or external influences (promoting situated cognition) play important roles in this enrichment model[14].

Similarly, Martin-Hansen (2002) described various levels of inquiry by way of a continuum. At one end, this continuum identifies a more teacher-centered 'structured inquiry' model as opposed to a more student-centered 'open inquiry' model. 'Guided inquiry', a combination of both, is found in the center. Defining elements are based on the genesis of the research question to be investigated. In structured inquiry, the problem is identified by the instructor, whereby the inquiry activities have predictable outcomes (cookbook style). In open inquiry, students determine the question for study as well as problem solve. Guided inquiry activities allow students freedom to design methodology and analytical techniques based on a given research question[10].

METHODS

The application of nanoscience in the non-traditional classroom was accomplished by way of individual and group activities. Exploration of Types I, II and III enrichment activities, as previously described, promote inquiry on a variety of levels. Examples of structured inquiry (associated with the Type I model) include experiences involving guest speakers and field trips. Students in the SRP visited a number of research labs affiliated with the CRISP a NSF funded Materials Research Science and Engineering Center (MRSEC) at Yale University. Prior to the field trip, students were given background information (via internet) pertaining to each of the lab groups, professors and research facilities. On a separate occasion, Prof. Mark A. Reed, Yale Applied Physics Laboratory, gave a presentation at the high school related to global implications of nano- and materials science research. In addition, several students visited the CRISP Open House held at Southern Connecticut State University (SCSU) and learned about advanced microscopy tools and techniques. With the exception of the field trip, students summarized their experiences in the form of a written report.

Enrichment activities that promoted guided inquiry (Type II activities) included the use of kits made available through CRISP. These NanoSense materials were developed by SRI International, with support from the NSF[1]. The purpose of these NanoSense projects was to increase awareness of current environmental issues and green technology. Upon completion of the kit (a Type I, hands-on, step by step process), students shared their findings with other classmates in the form of a lesson plan (Type II, guided inquiry activity) – one of many lesson plans that were constructed by each student as part of a group activity. These groups were designed around the content of each kit and included water filtration and organic solar cell.

To expand, the "NanoSense" project involving the construction of a nanocrystalline solar cell not only exposed students to a hands-on demonstration of current technology available to possibly combat global warming, but also educated students on environmental matters of global importance. This group created a working nanocrystalline solar cell – a more cost effective alternative to traditional silicon cells which mimicked the natural process of photosynthesis in plants. Students worked to replicate this process by using nanoparticles of titanium dioxide and certain organic dyes to create a flow of electrons. Through this hands-on research, students were familiarized with the current energy crisis and the promise of solar energy as an energy alternative. Students attempted to construct a working solar cell with inexpensive nano-materials in a non-traditional science classroom setting. Construction of this solar cell took approximately three weeks outside of the normal classroom setting. Significant time was invested in the troublesome task of annealing the titanium dioxide suspension on the conductive lenses of glass as well as the messy job of filtering gobs of raspberry pulp. Three completed solar cells were the fruit of each group's labor. After completing the nanocrystalline solar cell, students measured the energy produced using an ohm meter.

Another example of a Type II – guided inquiry (with components of Type III – open inquiry) enrichment activity involved the design and assembly of a photolithography kit. Photolithography is a process that is used in the integrated circuit board industry which uses light to transfer a geometric pattern from a mask onto a light-sensitive chemical, known as photoresist, applied onto a substrate such as a silicon wafer. The mask acts to selectively block light exposure and allow portions of the photoresist to become soluble to the photoresist developer solution while unexposed parts remain insoluble. A series of chemical treatments is then used to remove soluble parts of the photoresist and chemically etch the pattern onto the material underneath the photoresist, a conductive barrier layer. This process is significant in microfabrication because it allows for the selective removal of small parts of thin films and affords extreme precision in the shape and size of the objects it creates[11].

A team research opportunity was created after preliminary studies by their instructor were performed during the summer of 2008 through a Research Experience for Teachers (RET) sponsored by CRISP. The summer research took place at the Applied Physics Laboratory at Yale University under the supervision of Prof. Mark A. Reed and Prof. Ryan Munden. Three students attempted to duplicate the costly industrial technique in a safe, economical and educationally beneficial manner in the high school classroom. The purpose of this activity was to gain insight and appreciation for nanotechnology applications through an inquiry based activity corresponding with the classroom curriculum. Minor modifications to the original industrial procedure were made in order to simplify the process. For example, the high school version of the photolithographic process excludes the metal deposition layer or considers it an option for future work. Also, the spin-on technique for applying photoresist to the wafer typically used in the industrial setting was replaced by a spray-on version. Though photolithography is widely used in the construction of complex integrated circuits, the methodology can be applicable in high school science courses such as chemistry and physics. This investigation is currently ongoing in an attempt to gain successful image transfer in the high school.

The SRP offers a unique opportunity for all students to pursue Type III enrichment by conducting authentic scientific research primarily at a laboratory or academic research institution. Research projects are original and the brainchild of the student, and are performed under the tutelage of a mentor. The program combines novel research with experience using scientific method and community, supporting a common passion for a specific research field.

One example of a materials science based, authentic, open inquiry project was conducted by a student in the year 2009. In the experiment, *Determining Water Quality with Energy Dispersive Spectroscopy*, the Energy Dispersive Spectrometer (EDS) component of a Scanning Electron Microscope (SEM) was used to classify the elemental composition of water sediment from many different sources using relative percentages[12]. The water was prepared for analysis using a novel method designed by the student. Since the student identified a research question with regards to water quality; conducted original research with a mentor; adapted established methods to the specific scenario at hand; and created a new method of preparation, this project is exemplary of Type III – open inquiry enrichment. It also provides evidence that both SEM and EDS could be utilized by a high school student to conduct valid scientific research. The project allowed the student to become familiar with the scientific community, the field of physical sciences, and conducting original research.

Another nanoscience based Type III student-centered open inquiry study entitled *"Optimization of Surface Modification of PLGA Nanoparticles Using Avidin-Lipid Bioconjugates"* involved poly (lactic-*co*-glycolic acid), or PLGA, nanoparticles, modified to display avidin on the particle surface. These particles were used as vehicles for targeted drug and antigen delivery. The engineering objective of this study was to improve the density and stability of surface-presented avidin on PLGA nanoparticles, thus improving the efficiency of drug delivery. Particles were made using avidin-lipid bioconjugates with lipid chains of four differing hydrophobicities. Nanoparticles were analyzed with the SEM to measure the average size. Surface avidin presentation was quantified with bicinchoninic acid (BCA) protein assay. Avidin loss/stability was measured in an *in vitro* controlled release study. The student conducted research in a laboratory at Yale University's School of Biomedical Engineering. Though conducted under the supervision of a mentor, the project was largely designed and executed independently by the student.

CONCLUSIONS AND FUTURE WORK

The non-traditional, self-regulated nature of the SRP offers unique student learning opportunities in nano- and materials science, promoting creativity as defined by Firenstien and Treffinger (1989)[13]. The multidisciplinary content offered through the SRP appeals to a diverse group of learners by increasing student motivation and connections to an authentic research question. Interest in this elective program has steadily increased since it's inception four years ago. Initially, the program began with one section of science research (16 students) and doubled in size the following year (2 sections, 32 students). The third year boasted 3 sections (36 students). Currently, 42 students are enrolled in one of four SRP classes and an additional 2 students are conducting independent research studies due to a scheduling conflict. Within these classes, structured, guided and open inquiry enrichment activities (Types I, II and III) related to nano- and materials science are available for individual students, groups or research teams[2,3,14]. Mentors from industrial and academic settings promote student collaboration, situated cognition and social constructivism[7,9]. Collaborations with NSF Centers such as CRISP and availability of NanoSense projects encourage the investigation of global issues and promote student exposure to cutting edge technology[1]. RET's are beneficial to student learning by providing teachers with research experience that can be modeled, shared and/or elaborated upon in the classroom. One example of this would be the team photolithography project, originating from the summer 2008 CRISP RET.

Future work includes continued development of the photolithography kit for high school use by the SRP research team in conjunction with the CRISP educational outreach group. Correlation studies involving early exposure to SRL, inquiry and situated cognition to improved overall performance on the post-secondary level would prove beneficial. If a strong correlation exists, encouragement of a new educational paradigm based on the benefits of open inquiry (Type III) enrichment activities in the high school would be in order. Also, regarding benefits to general instruction, an increase in the availability of kits, curricular modules and supplemental activities promoting nano- and materials science would support higher-level thinking, problem finding and solving, long term memory and creativity as evidenced in the literature.

ACKNOWLEDGEMENTS

This research and the educational programs described are supported by NSF Grant MRSEC DMR05-20495. We also gratefully acknowledge the support of the CRISP Nanocharacterization Facility -- specifically Prof. Christine Broadbridge, Prof. Mark Reed Group, Monica Sawiki, Jacquelyn McGuinness Garofano; Amity Senior High School Board of Education, Superintendent, Administration and SRP student body; Yale University Lab Group under PI Dr. Tarek Fahmy; High School Educator/Research Instructor Dr. Frank LaBanca.

REFERENCES

1.   NanoSense materials were developed by SRI International, with support from the National Science Foundation under Grant No. ESI-0426319.
2.   Reeves, Todd D. "Toward a treatment effect of an intervention to foster self-regulated learning (SRL); an application of the Rasch Model." Thesis. Graduate School, University at Buffalo, State University of NewYork, 2009. Print.
3.   Renzulli, J.S. (1977). *The enrichment triad model: A guide for developing defensible programs for the gifted and talented.* Mansfield Center, CT: Creative Learning Press.
4.   Pavlica, Dr. Robert. "Replicating a Successful Science Research Program." *Journal of Secondary Gifted Education* 15.4 (2004): 148-54. Print.
5.   Iran-Nejad, A. (1990). Active and dynamic self-regulation of learning processes. *Review of Educational Research, 60*(4), 537-602.
6.   Puustinen, M., & Pulkkinen, L. (2001). Models of self-regulated learning: A review. *Scandinavian Journal of Educational Research, 45,* 269-286.
7.   LaBanca, F. (2008). Impact of problem finding on the quality of authentic open inquiry science research projects. Unpublished doctoral dissertation. Danbury, CT: Western Connecticut State University.
8.   Roth, W-M. & Roychoudhury, A. (1993). The development of science process skills in authentic contexts. *Journal of Research in Science Teaching, 30,* 127-152.
9.   Frawley, William. *Vygotsky and cognitive science language and the unification of the social and computational mind.* Cambridge, Mass: Harvard UP, 1997. Print.
10.   Martin-Hansen, L. (2002). Defining inquiry. *The Science Teacher, 69,* 34-37.
11.   Jaeger, C. Richard. Introduction to Microelectric Fabrication, Second Edition, Volume V.: Prentice Hall, 1996.
12.   Van Grieken, R., A. Markowicz, and Sz. Török. "Energy-Dispersive X-Ray Spectrometry: Present State and Trends. "Fresenius' Journal of Analytical Chemistry 324.8 (1986): 825-831.
13.   Firestien, R.L., & Treffinger, D.J. (1989). Update: Guidelines for effective facilitation of creative problem solving. *The Gifted Child Today, 12,* 44-47.
14.   Renzulli, J.S. & Reis, S. M. (1986). The enrichment triad/revolving door model: A schoolwide plan for the development of creative productivity. In J.S. Renzulli (Ed.), *Systems and models for developing programs for the gifted and talented.* Mansfield Center, CT: Creative Learning Press.

102

Mater. Res. Soc. Symp. Proc. Vol. 1233 © 2010 Materials Research Society   1233-PP07-02

Materials Science as a High School Capstone Course for the Physics First Curriculum

Nathan A. Unterman
Science Department, Glenbrook North High School, 2300 Shermer Road, Northbrook, Illinois 60062-6700, U.S.A.;
National Center for Learning and Teaching in Nanoscale Science and Engineering, 1801 Maple, Northwestern University, Evanston, Illinois 60201-3149, U.S.A.

ABSTRACT
By changing the high school science curriculum from Freshman Science, Biology, Chemistry, and Physics (BCP); to Physics, Chemistry, and Biology (PCB), we have an opportunity to create a new Senior level science elective. The entire high school science core curriculum has been reviewed and parts rewritten to create a coherent, integrated program based on common themes such as energy, particulate nature of matter, and forces. Nanoconcepts including size and scale and surface area to volume ratio are integrated where appropriate. In our school, we began PCB during the 2008-2009 academic year. In anticipation of these students becoming upperclassmen, a capstone elective course of Materials Science has been developed based on scientific models and literacies shaped in the PCB course sequence. Deployment of this new model-centered course is set for the 2010-2011 school year.

INTRODUCTION
Glenbrook North High School is changing to a PCB curriculum. Many other secondary schools are in the process of this change. Instead of just rearranging current offerings, as has been done elsewhere, a complete review and overhaul of core curricular offerings was made, including ways to identify and address student naïve notions of science concepts. Only Physics Education Research Groups have much to offer precollege schools in this area, and some Chemistry Education Research Groups are beginning to emerge with similar objectives. With the completion of the Physics and Chemistry rewrites, a concern for a senior-year core science course was raised. Materials Science was chosen as a capstone course for a coherently developed four-year high school science sequence. Currently, there is no formally produced Materials Science course in secondary education. This Materials Science curriculum was written as a reasoned continuation of the PCB sequence and continues to confront student misconceptions. This Materials Science course is not an independent course, but is a continuation of a curriculum that is a result of a complete, integrated rewrite of high school science. Student misconceptions identified in the Nanoscience Concept Inventory and Materials Science Concept Inventory (the MSCI is in development) played an important part in the creation of this course. What is presented here is just the Materials Science part of this entire secondary science curriculum revision.

The progress of civilization is an integral part of the history of Man and his materials – the Stone Age, the Iron Age, and today's Age of Silicon. Materials determine the technologies that provide protection, communication, information, construction, mechanization, agriculture, energy, transportation, and health. This laboratory-based course will introduce students to the production, processing, behavior, selection, and uses of six classes of materials: metals, ceramics, polymers, composites, semiconductors, and biomaterials. Knowing why glass shatters, steel is tough, rubber stretches and recovers, nylon can be drawn, and how nano gold is different from bulk gold, makes possible the selection of materials for enormously different applications. Studying large failures such as the Titanic's sinking and bridges collapsing, and smaller failures like light bulbs burning out or clothing staining are important parts of this course. Special attention is given to nanoscale materials and devices because of their potential for defining the next generation of important materials and machines. A student generated design project using science and engineering principles will culminate this yearlong course.

THEORY
The Need for Curricular Change
Since materials science and nanoscience have become more important within the research community and is socially relevant within the media, it has become necessary to evaluate how well secondary science education is preparing students to study such topics. Although most current curricula at the secondary level do not explicitly include materials science, the National Center for Learning and Teaching Nanoscale Science and Engineering (NCLT), SRI International, and other groups[1,2] have identified the Big Ideas of nanoscience and materials science, and have begun to link the underlying concepts to current science education benchmarks[3,4]. Vanessa Barker, University of London, has compiled a list of specific areas where students demonstrably lack useful understanding in the sciences, including states of matter, particle theory, changes of state, physical and chemical change, open and closed systems, and bonding. An enormous body of research shows that many students aged 11-18 are likely to have misconceptions in these areas. As designed in the implemented parts of the PCBM curricula (physics and chemistry, to date), many of these misconceptions are being systematically addressed. It is believed that coherent curricular design with sequenced courses leading to a materials science capstone course will nurture an improved understanding of a common science language and stress the interdisciplinary nature of all science concepts. This, in turn, will increase scientific literacy and support science learning in higher education.

When addressing materials science, nanoscience, and nanotechnology as part of public outreach, the National Science Foundation states, "The systematic control of matter at the nanoscale has the potential to yield revolutionary technologies for electronics, medicine, aeronautics, the environment, manufacturing, and homeland security. Because nanotechnology is expected to bring profound economic and social impacts over the coming decade, leadership in nanotechnology development will be crucial to future U.S. competitiveness in the global economy."[5] The National Science Foundation forecasts that by 2015, newly derived nano-based products and technologies will generate over two million jobs worldwide, with estimated production costs approaching $1 trillion[6]. There is clear need to assure that students have familiarity with nano concepts in order to make informed socioeconomic decisions about the impact of this technology.

DISCUSSION
Rationale For Addressing The Need Through A Curricular Change
The design of this fourth year laboratory science course carefully considers the Physics and Chemistry curricula. In addition, a brief review of Biology was made. Looking at the literacies addressed in these courses, a capstone Materials Science curriculum has been developed to include science literacies that create a more robust and complete flow for the secondary science curriculum. Currently, most high school science elective programs (non-advanced placement courses) consist of just one year-long, stand alone laboratory-based course. Often this is a self-

contained Earth Science or Astronomy course. Materials Science will give students an additional choice in laboratory-based science.

Since Materials Science was developed to complement the PCB curriculum, it will specifically extend this course sequence flow. For students in the Biology-Chemistry-Physics (BCP) sequence, there will be some overlap, since the BCP sequence has not been reviewed for coherence. Materials Science requires the completion of physics, chemistry, and biology, and is optimized for the PCB sequence while allowing for BCP students to participate. Although there will be similar engineering and materials themes paralleled in Applied Arts (metals, plastics, woods, etc.), this course explores materials from a science perspective with engineering applications rather than engineered materials. Materials science will complement the Applied Arts electives.

Student Profile
The target student profile will range from Regular to Honor students interested in a fourth-year laboratory science course who do not want to take an advanced placement level course. The class is designed to meet the needs of a variety of learners. At the Honor level, the basic curricular design is modified to include more advanced laboratory activities with a higher level of analysis, greater graphical emphasis, and use of online databases.

Scientific Literacies for Materials Science
Working definition of literacy: Having knowledge or skill in a discipline characterized by lucidity and ease of discussion.
- The universe is predictable and "regular"
- Conservation of Energy and the Laws of Thermodynamics, Equilibrium, and Self Assembly
- Matter is made of atoms
- Electromagnetism
- The properties of materials depend on the identity, arrangement, size, surface area to volume ratios, and binding of the atoms of which they are made.
- Science is an ever-changing process where questions are answered by interpreting repeated measurements in a systematic investigation.
- Waves and wave mechanics.

Technology Skills
- Use spreadsheets to manipulate data and graph relationships.
- Use appropriate data acquisition equipment for collection and analysis of measurements.
- Locate, interact, and participate with course information utilizing various Internet-Based applications.

Models in Materials Science
A scientific model is a construct composed of narrative, diagrammatic, pictorial, graphical, and mathematical representations that can be used to test and predict a phenomenon. It must have coherence and self consistency.
- Crystalline nature of solids and bound states.
- Energy transfers and crystal structure.
- Metallic bond structure and behavior.

- Ceramic bond structure and behavior.
- Polymer bond structure and behavior.
- Combining materials: Composite behavior.
- Energy and fields with crystal defects.
- Size dependent behaviors.
- Waves in one dimension.
- Design process.

Big Ideas for Materials Science
Broad concepts that are used across the entire subdiscipline.
- Particulate nature of matter
- Energy and energy transfer
- Bound states, phases
- Atomic Structure, Quantum Mechanics (electron orbital structure)
- Forces and Fields
- Size and scale
- Measurement, tools, and techniques
- Defects; both wanted and unwanted.
- Surface area to volume ratios.
- Design process

Primary Cultural Tenets in Science
- Safety
- Science is practiced by a community that includes collaboration and peer review.
- Science is an ever-changing process where questions are answered by interpreting re-
  peated measurements in a systematic investigation.
- The scientific community extends to resources and contacts in the digital realm.
- Science is the search for the fundamental properties of nature; engineering is the applica-
  tion of these properties to structure, processing, and performance.
- Use Internet search engines to locate valid course information and Materials Science data
  bases.

Course Curriculum
The Materials Science course was developed using the evidence based design of Mislevy[7] and
constructing measures design of Wilson[8]. The specifics of these documents are too long to be
presented here, but are available from the author[9] upon request. The design includes literacy,
scientific model; fundamental concept, claim, evidence, task, instruction; and unpacking. A brief
survey of content follows.
I. Introduction
II. Crystalline Nature of Matter
    A. Bonding
    B. Crystals
    C. Amorphous solids
    D. Polymorphism
III. Electronic Applications

A. Resistors
B. Doping, semiconductors, point defects
C. Conduction bands, PN circuits, diodes
D. Dielectrics, capacitors
E. Manufacturing integrated circuits (mask, vapor deposition, etc.)
IV. Mechanical Properties of Metals
A. Strength
B. Stress – strain
C. Dislocations – putting them in, taking them out.
D. Alloys, strengthening alloys, solid solution.
E. Thermal effects (specific heats, phase diagrams, lever rule, lamella, etc)
V. Ceramics
A. Structure
B. Properties
C. Defects
D. Processing/Fabrication
E. Applications
VI. Size-Dependent Properties
A. Size and scale
B. Surface area to volume ratio
C. Bulk versus nanoscale properties
D. Self assembly, dominant forces
E. Nanofabrication
F. Probes/tools AFM, MFM, SEM
VII. Polymers
A. Polymer molecules
B. Thermoplastic and thermosetting
C. Copolymers
D. Defects
E. Mechanical behavior
F. Deformation and strengthening
G. Crystallization, melting, glass transitions
H. Rubber, elastomers
VIII. Magnetic and Optical Properties
A. Electrons and light
B. Color without pigment
C. Refraction
D. Photonics
E. Curie Point
IX. Biomaterials
X. Composites
XI. Environmental and Social Issues
A. Recycling
B. New materials and biological interactions
C. Natural resources
D. Energy

E. Water
XII. Design project

Assessment
Pre and post testing using the Materials Science Concept Inventory (Arizona State and Ohio State), the Nanoscience Concept Inventory[10], and the TOSRA will be used to evaluate the effectiveness of student learning and attitudes. Additionally, laboratory exercises, homework, and summative and formative tests will be evaluated to track student understanding. Student misconceptions and naïve notions will be monitored and recorded for the development of classroom activities in order to confront the misconceptions and naïve notions.

CONCLUSIONS
The PCBM curricular change makes Physics and Chemistry service courses for Biology. Together, all of these courses lay a foundation for a capstone course in Materials Science. In developing new learning trajectories for all core science courses, Materials Science matches literacy objectives in creating a capstone course that unifies the Big Ideas in science. Materials have been an integral part of human history, and its study is important for defense, economic stability and development, and to have an informed citizenry. Through careful design and assessments, we are executing this course sequence and are collecting measures of the impact of this new sequence on student learning and attitudes in secondary science.

ACKNOWLEDGMENTS
Dr. R. P. H. Chang, Northwestern University Materials Research Institute, Evanston, IL
Dr. Michael Riggle, Northfield Township District 225 Schools, Glenview, IL
National Science Foundation
National Center for Learning and Teaching Nanoscale Science and Engineering

REFERENCES
1. Workshop hosted by University of Michigan and SRI International, June 14-16, 2006, in Menlo Park, CA
2. Stevens, Shawn, Lee Ann Sutherland, Patricia Schank, and Joseph Krajcik, *The Big Ideas of Nanoscience*, http://hi-ce.org/PDFs/Big_Ideas_of_Nanoscience-20feb07.pdf February 2007. Retrieved 15.September.2008.
3. American Association for the Advancement of Science, *Benchmarks for Science Literacy*, New York: Oxford University Press, 1993.
4. National Research Council, *National Science Education Standards*, Washington, DC: National Academy Press, 1996.
5. http://www.nsf.gov/eng/general/publicdoc/nanotechnology.jsp 12.August.2009
6. Roco, M.C., International Strategy for Nanotechnology Research and Development, *J. of Nanoparticle Research, Kluwer Academic Publ., Vol. 3, No. 5-6, pp. 353-360, 2001*
7. Mislevy, Robert J.; Almond, Russell G.; Lukas, Janice F., A Brief Introduction to Evidence-Centered Design. CSE Report 632, ERIC ED483399, US Department of Education, 2004.
8. Wilson, Mark, *Constructing Measures: Item Response Modeling Approach*, Lawrence Erlbaum Associates, Publishers. 2005.
9. nunterman@glenbrook.k12.il.us
10. Nanoscience Concept Inventory has been developed with the NCLT and further information is available from the author.

Mater. Res. Soc. Symp. Proc. Vol. 1233 © 2010 Materials Research Society   1233-PP11-04

Development and Delivery of an Online Graduate Certificate in Materials Characterization for Working Professionals

Pamela L. Dickrell[1] and Luisa A. Dempere[2]
[1]UF EDGE, College of Engineering, University of Florida, P.O. Box 116100, Gainesville, FL 32611, U.S.A.
[2]Major Analytical Instrumentation Center, Department of Materials Science & Engineering, University of Florida, P.O. Box 116400, Gainesville, FL 32611, U.S.A.

ABSTRACT

Within materials science and engineering industries there exists a need for continual professional development and lifelong learning. University materials science and engineering departments and materials related centers have highly qualified instructional faculty, and course management infrastructure that can be utilized to deliver needed continuing education to working professionals via distance learning. This work examines the development and first year delivery results of an online graduate certificate in modern materials characterization techniques for working scientists and engineers.

INTRODUCTION

Using industry demand as a driving force, a three course credited graduate-level certificate is created for place bound professional students with online, asynchronous delivery of course materials, without requiring participant travel. This certificate combines the practical expertise of a university based materials characterization instrumentation center with the instructional experience of graduate faculty to deliver course content relevant to practicing scientists and engineers who see the utilization of materials characterization in their workplace and desire a better working understanding of how techniques can be used to improve their research and development. In order to ensure success of distance delivery of a graduate certificate, three main organizations work together at the University of Florida. The first is the UF EDGE (Electronic Delivery of Graduate Engineering) Program, which houses all the distance learning infrastructure and services for the UF College of Engineering. The second is the Major Analytical Instrumentation Center (MAIC), which houses all the relevant materials characterization instrumentation and technical knowledge that is covered in the courses that are part of the certificate. The third group is the UF Department of Materials Science & Engineering which supports the offering of the certificate and whose faculty develop and deliver the curriculum covered in the online courses.

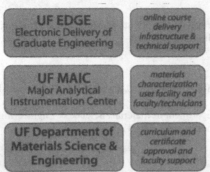

Figure 1. Three organizations within the University of Florida, College of Engineering and roles in the collaborative effort of an online certificate in Materials Characterization.

Online courses comprising the Materials Characterization certificate are all regular University of Florida graduate courses, captured electronically in specially equipped studio classrooms, with live audiences of graduate students concurrently enrolled in the same courses as the distance learning certificate students. Faculty are provided a range of instructional options including a regular chalkboard, prop camera, computer workstation, overhead projector, and a digital whiteboard, to ensure as much academic freedom as possible in structuring and delivering course materials. Each videoed lecture is posted online through the university course management system for distance students to view the same day as it is instructed live on campus, and distance students are expected to stay on track with campus students, completing the same homework assignments and exams within a one week window of campus students registered for the same course. Lecture videos remain online for both campus and distance students all semester to review before course quizzes or exams. The technical infrastructure for this asynchronous electronic delivery of UF engineering campus graduate courses is part of the UF EDGE program, which delivers courses from seven different UF engineering departments in this manner, to allow working professionals to complete individual courses, certificates, or full master's degrees in a distance, asynchronous environment.

THEORY

The intended audience for the certificate includes companies with research and development divisions, consultants dealing with characterization, and any scientists or engineers who see characterization used in their field and desire a better working understanding of the possibilities that materials characterization can provide.

Based on industry input, the online graduate Materials Characterization Certificate content is structured to balance both a relevant curriculum for working professionals, and the ability for successful certificate completion in one year or less. The first year offering of the certificate consisted of three graduate courses, offered one semester at a time; *Survey of Materials Characterization Techniques*, *Scanning Electron Microscopy and*

Microanalysis, and *X-Ray Diffraction*. Courses were chosen in areas of broad characterization practice and potential student impact, to encompass the widest user base for instruments, the greatest number of characterization applications possible in one year, and to focus on instruments that were the most affordable for research and development companies to have in-house.

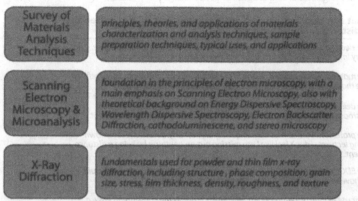

Figure 2. Structure and course descriptions for the first year offering of the online Materials Characterization Certificate.

Special considerations were taken in both the structuring of courses and curriculum modifications needed to existing graduate courses in materials characterization for relevant practical education of professional distance students, included introducing more practical examples of instrument uses and limitations, and in-class video demonstrations of techniques and instrumentation components.

DISCUSSION

Quantitatively, a comparator of performance between distance learning students and campus students is the grades earned in the courses. At the time of completion of this document grades were only available from two of the courses. Additionally, there were not a statistically significant number of distance student grades available for the courses on this first year offering of the certificate, but the preliminary average grade point average (on a 4.0 scale) do provide some initial data. In *Survey of Materials Analysis Techniques*, the distance student average was a 3.6 (5 students) and the campus student average was 3.4 (24 students). In *Scanning Electron Microscopy and Microanalysis*, the distance student average was a 3.1 (7 students) and the campus student average was 3.4 (41 students). With additional years of data from these certificate courses, these averages will begin to provide more statistically significant data. A preliminary evaluation of these first numbers shows that both distance and campus student performed academically similar to each other in these two courses.

A qualitative measure of the impact of the online Materials Characterization certificate is participant perspectives on their education gained, relevance of course materials and potential impact the certification will have on their current and future career goals. Three main questions, examining the online delivery, course content, and potential desire for hands-on training were surveyed of first year students. Student survey responses are edited for grammar and length for proceedings publication.

What was your impression on the delivery of course materials to distance/online students?

"The electronic delivery of course lectures and notes allows a lot of flexibility. That is very important in helping to juggle my work/travel schedule and family..."

"The material was presented clearly in the lecture videos, and it was very useful to have those videos to reference when I wanted to review certain parts..."

"The delivery of course material is exactly the same as being in the physical classroom with the exception of not being able to ask a question on the spot..."

"The professors were always able to quickly answer any questions I took down during lectures. I don't feel like I am missing out on anything by being a distance student..."

"The EDGE program opened up an educational opportunity for me that would not have been possible if I had to attend classes on campus..."

"Their use of high resolution video files makes reading the chalk board easy, a common problem with most online courses. Also, the files are posted online almost immediately after class ends and not the following day. The website where the course information is stored is easy to use and allows classes to be downloaded quickly."

Figure 3. Student survey results on online delivery format.

Were there any specific topics covered in the courses that you have found particularly useful or relevant in your job?

"The SEM and Survey courses have helped me have more effective interactions with the other researchers with whom I work..."

"At my current job I use the SEM almost daily, but I never had a course on the details of how it works and why. I learned both of these and my ability to operate the SEM has been enhanced as a result..."

"We have an SEM machine in our lab and as we were learning the theory I could discuss it with my coworkers. I now understand the limitations of the instrument, what I can ask the lab to do. This class was very relevant for my career..."

"The SEM course and Survey of material analysis techniques were a great overview of laboratory techniques commonly used. It is a good introduction for those ready to begin to working with any of the instruments..."

"The Survey of Materials Analysis Techniques course provided an adequate overview, without becoming overly basic, of the wide variety of materials characterization techniques available, many of which I was only vaguely aware prior to taking the course. Not only were the techniques themselves introduced, but also how they could be applied, including the advantages and disadvantages of using one technique over another..."

Figure 4. Student survey results on curriculum relevance.

If there were a three day weekend hands-on laboratory but it required traveling to campus, would this be something you would be interested in ?

"Maybe. A hands-on experience that correlated to the course material would be fantastic for filling in any mental gaps and also for solidifying the practicality of the material. However, I'm not certain how many campus visits, if any, my work would be willing to sponsor. It would depend on the financial decisions concerning this type of training/travel that my employer would make..."

"It might be intellectually interesting and I could see the value of some hands on work, but overall I wouldn't be interested. The reason why is that I live about 4 hours from campus, which means I lose a whole day of travel to attend the lab and would probably be burdened with hotel expenses..."

"YES. To experience and learn on equipment is the BEST way to retain the knowledge and apply it later (not necessarily on the equipment but within discussion or engineering requirements). I would love to have used half the equipment talked about within courses..."

"Since operation of the SEM is so hands-on, I would have been very interested in traveling to campus for one weekend during the semester to learn in the lab. Some concepts are better learned by doing them yourself rather than having someone tell you how to do it..."

"Perhaps a filmed "lab" where we could watch the laboratory classes since we cannot participate with hands on would be helpful. I would like to watch sample preparation for the TEM and SEM, AFM operation, EBSD prep and data collection, any of the techniques if we could watch the samples be prepared, analyzed and data interpreted from start to finish..."

Figure 5. Student survey results on potential desire for supplemental hands-on training on materials characterization techniques.

Student responses to survey questions indicated overall satisfaction with online delivery format of course lecture materials, course curriculum content, and showed that even if it did require in-person attendance, some working professional students would be willing to travel to campus for a three day hands on session of instrumentation techniques covered in the online graduate Materials Characterization certificate. Student suggestions for improvements in the program included the possibility of adding an optional hands-on weekend on some of the materials characterization instrumentation, more options for online characterization course offerings, and the addition of instructional technologies that would allow for more personal interaction with course instructors.

CONCLUSIONS

The first year delivery on an online graduate certificate in materials characterization for working professionals did deliver a one year option for better working understanding of a broad range of characterization techniques used in industry research and development.

Plans for subsequent year offering of materials characterization certificate courses includes the introduction of more interactive methods for distance students to experience characterization techniques. The first instructional supplemental method that will be introduced is a video 'lab' session, where the complete processes, including sample preparation, substrate or holder mounting, loading into instrument chambers, and operation of characterization instruments is captured for online video delivery of laboratory techniques. The second item to be introduced will be interactive, remote user software for the Scanning and Transmission Electron Microscopes, which will allow distance students to take online computer control of the focus, motion, imaging, and magnification aspects of the microscopes, providing some 'hands-on' learning while not requiring campus attendance.

Additional characterization courses will be introduced in future years of the materials characterization certificate. The Survey of Materials Analysis Techniques course will remain the one core broad topic course required in the certificate. While Scanning Electron Microscopy and X-Ray Diffraction are widely used techniques, additional course options in Surface Analysis (X-ray Photoelectron Spectroscopy, Auger, Secondary Ion Mass Spectrometry), Transmission Electron Microscopy, and Surface Metrology (Atomic Force Microscopy, optical, interferometer) are being planned for rotational introduction as elective options to comprise a 3 course, 1 year graduate level online Materials Characterization Certificate.

Mater. Res. Soc. Symp. Proc. Vol. 1233 © 2010 Materials Research Society   1233-PP07-04

Design and Implementation of Professional Development Seminars in Coordination With Research Experience for Teachers (RET) and Focused on Professional Practices of Scientists and Engineers

Chelsey Simmons[1], Kaye Storm[2], Gary Lichtenstein[1] and Beth Pruitt[1]

[1]Mechanical Engineering and [2]Office of Science Outreach, Stanford University, Palo Alto, CA 94305, USA

ABSTRACT

Many programs promote professional development for teachers in laboratory settings. In fact, some research has shown these experiences can improve student achievement. However, it is unclear what aspect of the laboratory experience helps bring about this effect. In order to ensure all teachers participating in Stanford's Research Experiences for Teachers program received maximum benefit from the laboratory experience, supplementary seminars were delivered that emphasized a variety of skills and tasks required of career scientists and engineers. Teacher feedback indicates that participants found these seminars valuable and that they would prefer additional time for peer interaction and curriculum development.

INTRODUCTION

National Science Education Standards established by the National Research Council in 1996 suggest that science teachers "encourage and model the skills of scientific inquiry, as well as the curiosity, openness to new ideas and data, and skepticism that characterize science" [1]. Exposing students to this expansive representation of science is expected to improve their skills as technical workers and as thoughtful citizens. It is also thought to improve students' attitudes toward and perceptions of science, though the mechanism of altering each may differ.

Research suggests that teacher participation in laboratory-based professional development can help student achievement in a variety of measures [2],[3]. However, one comprehensive study suggests that teachers participating in the National Science Foundation's Research Experiences for Teachers (RET) may not actually be conducting hands-on research [4]. In order to ensure all teachers participating in Stanford University's RET program were aware of the variety of skills and characteristics possessed by career scientists and engineers regardless of their laboratory assignment, we developed and delivered supplementary seminars that highlighted and reinforced professional practices. These seminars were thematically organized around the following professional practices of scientists and engineers:

- Analyzing and synthesizing research literature, including planning experiments and writing proposals;
- Collaboration, specifically the skills required to navigate diverse backgrounds, distributed tasks, and individual goals but shared resources;
- Synthesizing data and communicating results, including formal and informal mechanisms.

By creating approximations of these professional practices for teachers to engage in, we hope to create a more authentic experience for all teacher participants that can then be brought back to the classroom.

THEORY

As mentioned above, national expectations require teachers to convey an expansive view of science and scientific inquiry to their students. Many studies since have sought to explore ways that training and development programs cultivate habits in teachers that promote a broad and inclusive depiction of science beyond mandated content. Richardson and Simmons established a widely-used survey, the Teacher Pedagogical Philosophy Inventory, to examine teachers' beliefs about their training and subsequent teaching philosophy [5]. Studies done with this instrument show that discontinuities exist between teachers' understanding of the skills characteristic of scientific inquiry (embodied in "professional practice" above) and their ability to promote those skills in the classroom [6].

Recent studies suggest professional development in a laboratory setting, such as the RET program, may be one way to overcome these discontinuities. Teacher professional development in laboratory skills has been found to advance students 44% higher in grade level using the National Assessment of Educational Progress, compared to students of teachers who had not received similar professional development [2]. Another study found a 10% improvement in student pass rates on the New York Reagents' Exam for students of teachers participating in Columbia University's RET program [3].

While data linking professional development in laboratory settings with student performance exist, few data sets help explain this connection. However, pedagogical theories offer some suggestion as to this mechanism. Grossman et al. suggest that without firsthand participation and orchestrated opportunities for application, teachers and other learners find it difficult to translate theory into practice [7]. They deemed these opportunities for application "approximations of practice."

The National Science Foundation instituted Research Experience for Teachers (RET) with the goal of "involving teachers in engineering research and helping them translate their research experiences and new knowledge of engineering into classroom activities" [8]. However, not all those who have participated in RET actually have been involved in research. While over 80% of teachers nationally worked on classroom plans and *observed* research activities during the RET program, only 50% collected or analyzed data to answer a research question [4]. Thus, only half of the teachers participating in the RET program nationally actually have a chance to perform approximations of practice, running the risk of inadequate transfer of knowledge into practice. At Stanford, most teachers are placed in laboratories with opportunities for hands-on research, but we are not exempt from the larger trend. Last summer, about 15% of our teachers were focused on curriculum and web-based learning projects instead of inquiry.

In order to ameliorate this potential problem, we designed and implemented an eight-week seminar series to enrich the RET program by scaffolding for teachers the translation of theory and observation into their teaching practice. The seminars are designed to engage teachers in approximations of practices they witness in Stanford science and engineering laboratories and consider ways to incorporate appropriate approximations of practice into courses they teach at the secondary level. By focusing on approximations of practice, the desired impact is two-fold. First, we hope that this pedagogy provides a more tangible framework for teachers to translate their laboratory experience into the classroom. Second, by engaging teachers in approximations of practice as a group, we hope to create a more authentic experience for the small subset of our teachers not engaged in first-hand inquiry in their respective laboratories. We also involve all teachers in a wide variety of practices like grant writing and poster presentations, which are not necessarily part of a standard inquiry experience.

RESULTS AND DISCUSSION

To articulate the approximations of practice we presented to the teachers, we have detailed below the activities and assignments carried out with the cohort of 21 teachers involved in Stanford's RET in the Summer of 2009.

Exposure to the laboratory setting and introduction to approximations of practice

To help teachers acclimate to the laboratory environment, they were provided with select chapters of a laboratory manual before they arrived on campus [9]. Additionally, we asked each mentor (typically a professor or graduate student) to provide us with background reading for the teacher. The mentors typically responded with articles from technical literature, which were forwarded to the teachers along with a worksheet providing tips on reading technical literature. While teachers indicated that it was overwhelming to receive assignments prior to their first day, they indicated that they particularly liked the readings from *At the Bench* [9].

In addition to discussing the readings during the first seminar, we also discussed the pedagogy of approximations of practice mentioned earlier. In order for teachers to relate this to their teaching, we conducted an interactive activity where teachers were asked to share approximations of practice they already use in their classrooms. Blank poster boards titled with one of the three professional practices were posted around the room. Teachers were asked to write activities they use that are approximations of these practices, as well as list problems they encounter while trying to complete these activities, as they rotated around the room. Teachers reported that they enjoyed the opportunity to share ideas with other teachers in this manner.

Analyzing and synthesizing research literature

To begin the discussion of reading research literature, we asked teachers to review a journal article and popular magazine article written about the same technology. This was used to instigate a discussion about the differences and similarities between the two genres and how each may or may not be appropriate for the classroom. Teachers were also asked to "reverse engineer" their thinking in answering questions about the papers. They realized that by simply asking personal questions about the content, e.g., "Would you want this technology in your home?," they could make the material relevant to the students. Teachers mentioned, though, that they would have liked to spend more time translating these ideas into actual curriculum components they could use in their lessons.

We also spoke with teachers about the role synthesizing literature plays in designing experiments and proposing new ideas, namely in grant writing. In order to provide an approximation of grant writing practices, teachers were given a mock grant assignment with guidelines to write a grant on whatever they desired. A few excited teachers outlined scientific grants related to their summer research, but more teachers took the practical route of planning activities for their classrooms. The seminar leaders provided extensive feedback to all teachers who participated and prizes were awarded for the most well written grants.

Collaboration

We created a cooking activity for the teachers to stress the importance of collaboration. The teachers were split into groups, and a Principal Investigator (PI) was designated (not from volunteers) for each team. The PIs were given a recipe that was written in French, and each group member was given the translations for only a certain portion of the recipe, thereby enforcing communication and teamwork. Furthermore, the teams were not given enough cookware to finish their recipes independently, thus they had to coordinate sharing equipment with other teams, much like a real-life laboratory. During the race to be the first team to finish cooking, many authentic conflicts were created. This provided a powerful, shared experience for teachers to begin discussions of how to modify classroom group work to constructively mimic some of these dynamics. In debriefing the session, we asked teachers what they would bring back to the classroom from the experience. They mentioned the value of appointing leadership roles, rather than always taking volunteers, and of having their different classes communicate with one another, perhaps providing each other data or working on separate aspects of related problems.

Another realistic aspect of collaboration we wanted to stress to the teachers is the diversity of personalities and skill sets present in a research setting. We asked teachers to interview their mentor about their background and experiences in the lab. We also spent time as a large group discussing the variety of roles witnessed in the laboratory. Teachers reported that this broadened their perspective on the role of a research scientist, which was a welcome result. Out of the diversity of skill sets we were hoping to highlight, we spent an entire session specifically on creativity. In a reflective activity, we asked teachers to consider the nature of the work they and their mentors were undertaking and whether it matched the work they ask their students to do in the classroom. The teachers then discussed in small groups how to incorporate more creativity and higher order thinking skills into their assignments.

Synthesizing data and communicating results

To paint a broader picture of the how scientific discoveries are communicated to society, the teachers heard from the Stanford Office of Technology Licensing. This presentation, along with personal anecdotes from one of the authors (C.S.) about her experience with patenting an invention, instigated much discussion. Teachers were particularly interested in how to encourage their students to pursue inventions and understand the patenting process, as well as laws that govern fair use of multimedia in the classroom.

During the last meeting of the summer, teachers were also asked to participate in a conference-like poster session as an approximation of practice related to data analysis and communication skills. Teachers viewed each other's posters, gaining exposure to research in other fields, and presented their own, gaining practice at communicating advanced scientific concepts. Preparing the poster also gave many teachers practice working with PowerPoint and other software packages they may not normally use.

Presentations from Office of Science Outreach and curriculum component with IISME

In keeping with the theme of communicating results of scientific inquiry, the Office of Science Outreach coordinates science-oriented presentations from faculty across the university. Many talks and field trips occur over the course of the summer, and teachers appreciate this exposure to research occurring across the university. Additionally, Stanford pairs with Industry Initiatives for Science and Math Education (IISME) to work with RET teachers over the summer. IISME requires teachers to develop a curriculum unit, an "Education Transfer Plan," that corresponds to California State Standards in their subject. IISME's support via a peer coach and online infrastructure were also important components of the teachers' experience, ensuring that their laboratory experience was transferred to the classroom via standards-based curriculum.

Feedback from returning teachers

Teachers who had participated in the Stanford program previous summers were asked by email about their 2009 experience that now included these supplemental seminars. Their responses are summarized in Table 1.

CONCLUSIONS

Overall, we were pleased with the results of our pilot experiment with supplemental seminars. As you can see in the representative feedback we received, teachers highly value time

Table 1. Sample responses from teachers who had been at Stanford for two RET summers with and without supplemental seminars. The top row lists a mixed comment, characteristic of the majority of the six responses, with suggestions for modifying future seminars. The second row lists the most positive review received and the third row, the most negative.

Do you feel the seminars offered in 2009 enhanced or detracted from your research compared to previous summer(s)? Please elaborate on your answer.	*Do you feel the seminars offered in 2009 enhanced or detracted from your teaching and curriculum development compared to previous summer(s)? Please elaborate on your answer.*	*Would you recommend these seminars be included as part of the Stanford-IISME program in the future? If so, why? If not, what is your hesitation?*
I think it was somewhat useful for the research, mainly in how I thought about my research lab. But then again, my research situation was also quite a bit different than my previous experience.	In the sense that they crowded out time to work on lesson plans with other teachers and share lessons we have on certain subjects, it detracted quite a bit. ...I never quite got to the point where I felt like I had fully formed ideas on how to integrate more "real" science into my classroom as actual lessons.	One idea I had was perhaps it could be every other week? And on the off weeks people could get together in subject areas and discuss ways they teach specific subjects? ...Also, if the seminars could focus a bit more on ways teachers could use them in the classroom, or ways to create lesson plans, that would make it more interesting and useful long-term.
I feel that overall the seminars enhanced my research compared to previous summer experiences. The content complemented my experiences in the lab and provided me with a more in-depth look into the different facets of research.	I feel that my teaching and curriculum development have been enhanced. Some of the content we covered reaffirmed why I have students engaged in a certain activities...Also, knowing about the hierarchy in the lab helped me to better identify individual roles during lab activities and the emphasis on reviewing research to make more informed decisions has helped me identify more effective teaching practices.	I would recommend the seminars, but I think they need to be somewhat different from year-to-year for fellows returning to their labs. Also, it would be beneficial to incorporate more time for teachers to discuss their own practice with other fellows and share ideas about how to integrate seminar content into classroom activities.
For the most part, the seminars detracted from my research. I learned little about teaching science and turning students on to learning.	Compared to previous summers, it was about the same in terms of value to my teaching.	I recommend the seminars be led more by teachers helping teachers. Sharing labs is, for me, a better approach. It gets right to the heart of what we are doing as teachers and opens our minds to how we can improve our curriculum.

to interact with one another and work on concrete lesson plans they can take back to the classroom. Transferring ideas back to the classroom remains the most salient concern of teachers, and we still believe we can promote this by focusing on approximations of professional practices that make explicit for science and math teachers the skills and abilities professional scientists and engineers rely on implicitly. With our supplemental seminars, group discussions help teachers to articulate the technical and interpersonal skills required of scientists and engineers. By writing grant proposals and creating conference posters, teachers have firsthand experience with the intricacies of planning experiments and communicating results. Furthermore, team activities reinforce the need for classroom experiences that incorporate collaboration and opportunities for creative thinking. The seminar series helps teachers acknowledge the complexities and 'non-linearity' of laboratory research and broaden teachers' notions of skills and abilities valued by career scientists and engineers. In the future, however, we hope to build more opportunities into the seminars for teachers to interact with one another about lesson plans and in-class labs.

ACKNOWLEDGMENTS

The authors would like to thank the teachers and mentor labs that participated in the program in 2009 as part of this work. Teacher stipends were funded from nine NSF Centers and Programs, NIH, HHMI, David and Lucile Packard Foundation, San Mateo County Board of Supervisors, Stanford University, and William and Flora Hewlett Foundation. C.S. acknowledges an NSF Graduate Research Fellowship and Stanford Cardiovascular Institute Smittcamp Fellowship. This material is based in part upon work supported by the National Science Foundation under Grant Numbers CBE-0735551, ECS-0449400 and EEC-0908516. Any opinions, findings, and conclusions or recommendations expressed in this material are those of the authors and do not necessarily reflect the views of the National Science Foundation.

REFERENCES

[1]   "National Science Education Standards," National Research Council: National Academies Press, 1996, p. 262.
[2]   H. Wenglinsky, "How Teaching Matters: Bringing the Classroom Back into Discussions of Teacher Quality," Policy Information Center, Educational Testing Service2000.
[3]   S. C. Silverstein, J. Dubner, et al., "Teachers' participation in research programs improves their students' achievement in science," *Science,* vol. 326, pp. 440-2, 2009.
[4]   S. H. Russell and M. P. Hancock, "Evaluation of the Research Experiences for Teachers (RET) Program: 2001-2006," SRI International2007.
[5]   P. Simmons, A. Emory, et al., "Beginning teachers: Beliefs and classroom actions," *Journal of Research in Science Teaching,* vol. 36, pp. 930-954, 1999.
[6]   S. L. Brown and C. T. Melear, "Investigation of Secondary Science Teachers: Beliefs and Practices after Authentic Inquiry-Based Experiences," *Journal of Research in Science Teaching,* vol. 43, pp. 938-962, 2006.
[7]   P. Grossman, C. Compton, et al., "Teaching practice: A cross-professional perspective," *Teachers College Record,* vol. 111, pp. 2055-2100, 2009.
[8]   "Research Experiences for Teachers (RET) in Engineering," National Science Foundation.
[9]   K. Barker, *At the Bench: A Laboratory Navigator.* Cold Spring Harbor: Cold Spring Harbor Laboratory Press, 2005.

Mater. Res. Soc. Symp. Proc. Vol. 1233 © 2010 Materials Research Society   1233-PP03-06

Building a Low Cost, Hands-on Learning Curriculum for Glass Science and Engineering Using Candy Glass

William R. Heffner[1] and Himanshu Jain[1]
[1]International Materials Institute for Glass, Lehigh University, 7 Asa Drive, Bethlehem, PA 18015, U.S.A.

ABSTRACT

We have developed a program to connect students, as well as the general public, with glass science in the modern world through a series of hands-on activities and learning experiences using sucrose based glass (a.k.a. hard candy). The scientific content of these experiments progresses systematically, providing an environment to develop an understanding of glassy materials within a framework of "active prolonged engagement" with the material. Most of the experiments can be assembled in a high school lab, or even in a home setting with minimal cost, and yet are appropriate for inclusion in an undergraduate materials lab. The cost is minimized by utilizing common, everyday materials and devices. Some of the activities included in our experiments include: synthesis, density, refractive index determination, glass transition, crystallization, kinetics of devitrification, thermal properties, etc. Temperature measurement, temperature control, and even automated data collection are part of the experience, providing an open path for the students to continue their own interesting and creative ideas.

I. INTRODUCTION

The 1983 publication of A Nation at Risk [1] identified the decline in the academic achievement of US students and the potential for failing to meet the national need for a competitive workforce. Since that time much social and political dialog has centered on the need to improve student achievement and interest in science, engineering and technology education in the US. Recently more attention has been brought to the significance of both hands-on learning and the informal educational experience to the total educational experience of both student and adult learners [2]. In response to this challenge, we have developed a program to connect students and the general public with glass science in our modern world through a series of hands-on activities and learning experiences.

Glass and glassy materials are important and ubiquitous materials in our everyday life. In fact, they are perhaps among the most common material in our everyday experience, from windows, doors, kitchen ware, eyeglasses, cameras, and insulation, not to mention the optical fibers empowering the information age. And yet, with this incredible body of experiential familiarity to relate to, students experience little or no introduction to this important material in any of their formal high school or college science training.

One of the problems in conducting any serious investigation of glass science for the younger students (especially in the high school or home setting) is the high temperatures required to make or form these materials, especially the commonly used oxide glasses, as well as the specialized equipment required to process these hard materials. However, a much lower temperature example of glass is found in the universally pleasant world of candies, the sugar glass, also known as hard candy. The sucrose-corn syrup-water system of candy glass mimics many aspects of commercial, soda-lime-silica glasses and these close analogies have been described in earlier papers [3, 4].

In this paper we describe a series of hands-on exercises for the student to experience both glass technology and glass science through explorations with candy glass. The experiments build on one another to provide a mini-curriculum of low-cost, hands-on activities designed to facilitate a rich experience of active prolonged engagement with glass science. The apparatus required for these experiments can be assembled from commonly available items from the home, hardware store or the high school lab and there is an emphasis on building one's own equipment (such as a light bulb sample heating oven). Our intent is to engage the students by allowing them to learn largely through discovery, building knowledge and interest through successive experiences around a common glassy material that they can make and modify as their ideas evolve. While the experiments in our series are simple enough for the students to do at home, they are quantitative in nature and allow students to explore "real glass science". Many are ideally suited for the student science project, while some are appropriate for a classroom demonstration (like the fiber drawing tower).

The materials in our candy glass curriculum are all available and distributed through NSF's International Materials Institute for New Functionality in Glass (IMI-NFG) website at http://www.lehigh.edu/imi/. The use of Internet allows us to include a wide range of materials to support the learning experience, including tutorials, videos, project descriptions, student presentations and even construction details. Likewise the website provides a means for reaching a large population of students and teachers, while providing a fast and flexible means to revise and add new content as it is developed.

II. BACKGROUND

While sugars and candy glass are not part of the traditional material science domain, they do certainly have a large and important place in food science and some of the best introductory and advanced material on sugar glasses can be found in their collections. McGee's very enjoyable classic, On Food & Cooking: The Science & Lore of the Kitchen [5] is highly recommended as a starting point for the student. For the more serious student looking for detailed information on sugars, there are two excellent up to date references from the food science community, one is Crystallization in Food by Hartel [6], and the other is Sucrose Properties and Applications edited by Mathlouthi and Reiser [7]. The latter are both available on Google books.

For the material scientist the behavior of the sucrose water system is best visualized through the binary phase diagram shown in figure 1. Data for the solubility as well as the freezing point depression for sucrose in water are available, from which the students could readily construct their own diagram [8]. Data for the glass transition temperature are likewise available [9]. The figure shown here has been prepared from these data using Excel and is in good agreement with the diagrams reported in other sources [10]. In our diagram we also include the boiling point data taken from the Food Industries Manual [11].

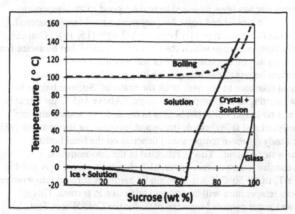

Figure 1. The sugar/water phase diagram for the sugar/water system with boiling point line included. Prepared from data referenced in the text.

The sucrose solubility vs. temperature curve (even included in some high school chemistry textbooks) defines the upper temperature of the two-phase region (crystal + solution). For temperatures above this line sugar crystals are unstable and will dissolve. Below the solubility line, sugar crystals are stable and can exist in equilibrium with the solution.

However, due to the high viscosity of the syrupy supersaturated solutions, crystals can be slow to form, especially in the absence of seed crystals or other nucleating sites. As a result high sugar content solutions have a distinct metastable region below the solubility curve. Supersaturation ratios as high as 1.2 are stable for a long period of time in sucrose/water systems, with spontaneous crystallization occurring at a ratio of about 1.3 [12, 13]. The supersaturation ratio, commonly referred to in the sugar literature, is defined as the ratio of the solution concentration (in wt. %) to the saturation concentration. Within the food industry it is well known that the addition of other sugars, in particular corn syrup (which contains glucose polymers), can greatly extend this metastable range [14]. In our experiments to follow we demonstrate this inhibition from corn syrup and utilize it in developing our own formulation for sugar glass.

The viscosity of the metastable syrup increases rapidly as the hot solution is cooled, eventually reaching the glass transition temperature. Once a glass is formed, the material is protected from subsequent crystallization due to the extremely high viscosity of this solid-like material.

In order to prepare candy glass, a mixture of sugar and water is first heated to dissolve the sugar crystals. For moderate sucrose content the solution will clarify prior to full boil as the sugar crystals dissolve. For high sugar content, the clearing point may occur after the solution begins to boil, as indicated in the phase diagram. By continuing to boil, the water content of the one phase solution can be gradually reduced, with a concomitant increase in the boiling temperature. Thus the boiling temperature can serve as a convenient measure of the water content and provides the simple monitor for determining when enough water has been removed to make a good room temperature glass. Boiling to 145-155°C is a typical end point for preparing a hard candy material. The moisture content of a hard candy is typically between 1-2 wt% [15, 16]. One can control the hardness by the temperature to which the mixture is cooked, but be aware that the temperature rises abruptly as you reach these lower water content region.

The solution temperature begins to rise abruptly as it approaches the requisite low water content and one must take care to not over cook the material. Sucrose begins to degrade (and turn yellow) starting at this temperature range. "Above 165°C the sugar is more than 99% sucrose and no longer boils, but begins to break down and caramelize" [17]. At 170°C it is hydrolyzed and splits into dextrose and levulose, or invert sugar [18].

The hardness of the candy (at room temperature) depends on the final boiling temperature (and thus the water content). This is reflected in the concentration dependence of the glass transition temperature curve, also shown on figure 1. A good hard candy should have a T_g of 40-50°C if it is to retain its hardness at room temperature. Candy with a T_g near room temperature will be soft and taffy like in texture. This relationship between T_g and water content forms the basis for the Cold Water Test also known as the Ball Test [19] used by old-time cooks to prepare candies long before thermometers were common in the kitchen. To perform this test, a small amount of the boiling solution would be dripped from a spoon into a dish of cold water, quenching the drops to near room temperature. Once cool the cook inspects the drippings to see if they are soft, hard or crack (brittle). Cooking to the soft ball state serves as the doneness test for fondants and fudge, while a hard ball is required for taffy and the hard crack for making good hard candies. For the interested reader McGee [19] provides a very clear discussion of this test and the various types of ball stages associated with different temperature ranges. We encourage all students to utilize these old time and highly intuitive monitors in conjunction with the solution temperature to develop a deeper understanding of the process.

While the two component sucrose-water system has the advantage of simplicity as well as a large body of well documented literature dating back to more than a hundred years, it has one serious limitation for actual candy glass making and student experiments. The sucrose-water only system is strongly susceptible to crystallization, very problematic for glass making. (Although controlled crystallization is actually required for the making of some confectionaries like fudge.) It is especially difficult to prepare good candy glass from a mixture of pure sucrose and water because these solutions have a relatively high rate of crystallization, especially at the low water content range required for candy glass. Frequently even stirring the hot solution can promote the onset of crystallization in simple sucrose-water recipes, ruining the preparation. As mentioned already, corn syrup has a large inhibitory effect on the crystallization rate and is almost always included with sucrose in the hard candy glass recipe. In one of the following sections we describe experiments which illustrate the effect of corn syrup to sucrose ratio on crystallization. For the bulk of our other experiments we have settled on a 2:1 ratio of sucrose to corn syrup as our basic recipe.

III. EXPERIMENTAL ACTIVITIES & MODULES
A. Synthesis of Candy Glass - Exploring the Properties and Applications of Glass

For our introductory activity in glass science we typically combine an interactive discussion of glass followed by showing students how to make their own sugar glass. Our version of the activity is typically delivered in a 90 minute period for a science camp or student/teacher workshop, but could easily be divided into two shorter periods to fit the class room schedule.

After a short discussion of the examples of glass in our everyday experience, we find it useful to encourage the students to list some properties of glass, based on their own experience. Typically the students will, with only minimal coaching, come up with a list which includes such attributes as hard, brittle, easily fractured, transparent, and flows on heating. By defining glass in terms of its common observable behavior, one can quickly establish the connection to hard candy as also representing a glass. The Jolly Rancher brand from Hershey Foods Corporation provides a convenient example of candy glass to include in the discussion. Focusing on properties provides an ideal spring board for a discussion of how glass is different from the other condensed matter phases (liquid and crystalline solid) and the structural differences associated with each of these states. Likewise the discussion of candy as a glass sets the stage and motivation for the making (synthesis) of hard candy glass.

There is an abundance of recipes for hard candy available from cook books or the internet. We also include a list of recipes in the glass making instructions. Our standard recipe for most experiments is a 2:1 (by wt) mixture of sucrose to liquid corn syrup with approximately 10% (by wt) water added at the start. The actual details of the recipe and the cooking procedure are provided on our website, with other relevant information. The cooking procedure involves monitoring the temperature with a low cost digital cooking thermometer as the glass making proceeds. The student is encouraged to note such phenomena as the onset of boiling, the temperature of complete dissolution (or clearing), as well as the emerging thickening ("stringiness") of the solution as the temperature rises and the water evaporates. The old fashioned techniques of testing for soft ball and hard crack, described in the previous section, are used to determine the progress of the preparation. By comparing such clear observables with the actual solution temperature the student begins to develop an experiential understanding of the glass forming process.

Once the hard crack state of the liquid is achieved (approximately 145-155°C), the syrup is removed from the heat, allowing a short rest period for the bubbles to escape. Next, the hot liquid is poured into molds, free form discs, sheets, test tube samples for experiments, or any number of shapes for consumption. To cap off this student activity we usually save some of the material for making candy glass fibers. Upon cooling to ~ 90°C the melt becomes ready for pulling long glass like fibers using wooden popsicle sticks. An on-line video shows a student pulling a candy glass fiber of more than 100 meters down a hallway [20]. In Section 3 below we describe a home built fiber drawing tower which provides a more quantitative extension to this initial exploration with fibers. After this first interactive lesson on glass and candy glass, the students are prepared to make their own candy glass at home and begin experimenting using some of the activities suggested below as a starting point.

B. Post Synthesis Activities

Once the students have learned to make their own sugar glass, a number of other interesting experiments can be explored. Two experiments that relate directly to the physics and chemistry classroom include density and refractive index measurements, while a fiber drawing tower provides an opportunity to explore an engineering based experience involving optimization of processing parameters.

1. Density

Density is an important property of glass, which often correlates well with other properties such as refractive index [19]. It is also a very basic property of matter introduced early into the science curriculum, so most young students will already have some familiarity with the concept. However, measuring the density of a water soluble, irregular shaped solid with standard laboratory resources provides an interesting challenge for the student to consider. The low cost method that we describe involves constructing a student-built pycnometer. The pycnometer is constructed from a small jar (e.g. 4 oz salsa jar) for constant volume, with a small hole drilled in the lid to allow for its filling with water to a fixed volume. The lid of the salsa jar is stiffened so that it would not deform (and thus change volume) by epoxying a steel washer to the top. This pycnometer can be constructed at very low cost (a few dollars at most for epoxy, washers and a jar of salsa or artichokes) and provides the students with a rich hands-on experience of actually constructing their own instrument.

Figure 2. Low cost apparatus for the measurement of density of irregular shaped solids, using Archimedes method.

To determine the density the pycnometer is weighed empty, filled with only water, and finally with glass of known weight and water. The weight of the glass divided by the weight of water displaced is the specific gravity. For density one must also include a factor for the density of tap water at room temperature (0.998). A simple triple beam centigram balance, available in most high school labs, will allow a density measurement to 4 significant figures.

Before tackling the candy glass density measurement, students should first explore the accuracy and repeatability of this method utilizing some standard oxide glass or other water insoluble material. By using a syringe to remove all air bubbles and to top off the water level to a fixed meniscus at the hole, we were able to achieve a standard deviation of 0.005 g/cm^3 for the density of pieces of a broken dish (oxide glass with nominal 1.68 g/cm^3 value). Achieving such accuracy requires careful attention to detail, including avoiding changes in room temperature. Once the accuracy of the method has been demonstrated for the water-insoluble materials, the student can tackle the density of their own candy glass. On the IMI-NFG website we include details for the density measurements made by one of our students on a collection of sugar glass samples made with varying ratio of sucrose to corn syrup. Our student investigator found no significant difference in density for the candies made to the same hard crack temperature over a wide range of sucrose to corn syrup ratio. By encouraging the student to consider some simple models for density of mixtures, the student should find this result quite plausible.

2. Refractive Index

Refractive index is another important property for a transparent, optical glass, and sugar glass provides an opportunity for the students to prepare their own samples as well as their own apparatus for measuring this basic parameter. We have examined two simple methods by which the student can measure the refractive index of sugar glasses made by themselves, using very basic apparatus. The simplest is based on Pfund's method [22], where only a laser pointer and a ruler are required to determine the refractive index of a thick, flat sample of a transparent material. If a transparent glass sample is illuminated with a laser pointer as shown in figure 3, then a change in reflected light will occur at the critical angle. The vanishing reflectivity at the critical angle produces a dark region of radius, r, around the center spot, if there is no air interface at the bottom surface. The refractive index and be easily calculated from the radius of the dark region and the thickness of the sample, h, using $n^2 = 1 + (2h/r)^2$. In the case where there is also an air interface on the bottom surface, such as a flat glass not adhering to the surface, then the light ring is observed in the center at the same condition.

The other, more accurate method, utilizes a student spectrometer and a prism of candy glass formed within a glass mold constructed by the student from glass slides. The minimum deviation method [23] allows the student to obtain the refractive index to four significant figures, while quantitatively exploring the nature of refraction. Each of these methods has been tested and described in detail by undergraduate students during IMI-NFG's summer REU programs at Lehigh University. Full presentations of their procedures and results are available on our website.

One of our students utilized the minimum deviation method to determine the refractive index of candy glasses over a wide range of sucrose to corn syrup ratio and found no significant variation of refractive index, consistent with the observed uniformity of densities. [We also include details on how to build spectrometer light sources of different wavelength from low cost LEDs, so that measurement of chromatic dispersion can also be explored by the student.]

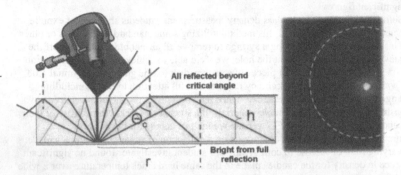

Figure 3. Basic concept of Pfund's method for measuring the refractive index of a transparent plate. Photo on right show darker region in the center of sugar glass slab.

Figure 4. Left, an REU student measures the refractive index of a candy glass prism. Right, empty and candy filled prism molds made from microscope slides by the student.

3. Fiber Drawing Tower

Early on in the preparation of candy glass the student has the opportunity to experience the unique ability of glass to form fibers from the melt once the temperature is within an appropriate range. The fiber drawing property is noticeably temperature sensitive, providing an excellent opportunity for the student to engage in an engineering experiment to optimize and control the process parameters required for drawing candy glass fibers. To facilitate this inquiry we have designed a sugar glass fiber drawing tower which mimics many of the components of an actual drawing tower used for fabricating fibers for optical communication, but only at a much lower temperature and made again from low cost parts. Our candy glass fiber drawing tower utilizes one or two halogen light bulbs to heat a rod of candy glass preform (fabricated by the student). The temperature of the sugar glass is monitored using a low cost thermocouple, allowing him or her to quantitatively explore the optimal conditions for fiber drawing. The take up spools have been made from empty, plastic, peanut butter jars. As with the other activities, details with construction sketches are included on our website.

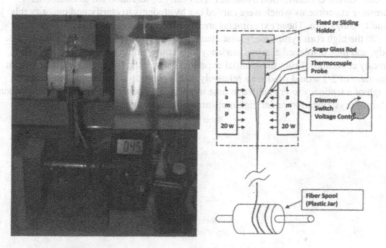

Figure 5. Photograph of our home-made fiber drawing tower for candy glass using two halogen lamps for the "furnace". A sketch of the components for measurement and control is shown on the right.

4. Crystallization

As mentioned in the Background section, pure sucrose has a very strong tendency to crystallize in concentrated aqueous solutions even at high temperatures, making it difficult to form a glass with sucrose as the only sugar component. The addition of corn syrup greatly reduces the tendency to crystallize, which is why it is essential in most recipes, as well as ours. This addition of a third component provides an excellent opportunity for the student to explore the influence of the sugar to corn syrup ratio on the glass formation properties of this system. One of our early experiments was to examine this dependence of glass stability on composition in the sucrose/corn syrup/water system. A pseudo-ternary phase diagram for this system is illustrated in figure 6 below. Samples with varying sucrose to corn syrup ratio were all heated until the solution was at 145° C and then poured into paper containers to cool. Black paper was used for the base, so that scattering from any crystal formation would be easier to photograph.

On the high sucrose, low water side, the boiling solutions are very prone to crystallization as the water is driven off during cooking, making it nearly impossible to form a glass in a sauce pan from an all sucrose mixture. At 80-90 % sucrose to corn syrup ratio it is possible to form a candy glass on cooling, but the resulting material is very prone to subsequent crystallization within less than a day. A 70% or lower sucrose to corn syrup ratio is recommended to avoid severe crystallization problems. An insert in figure 6 illustrates how these crystals begin to form at the surface of these glasses. This tendency to form surface crystals was observed to be far more rapid during periods of high humidity, an observation suggesting that atmospheric moisture has a considerable influence on the crystallization dynamics. This early observation led to additional interesting experiments which were carried out by students to clarify and quantify this humidity conjecture. These experiments are described in the next sections.

On the high corn syrup side, glasses tended to be much softer than the high sucrose candy, and they displayed a strong tendency to become gummy or tacky under high humidity conditions. From these initial experiments we established that sucrose to corn syrup ratio of 2/1 would produce a relatively stable glass on cooling, and yet retain enough of a tendency to crystallize in time to provide a convenient system for subsequent studies of crystallization under high moisture and elevated temperatures.

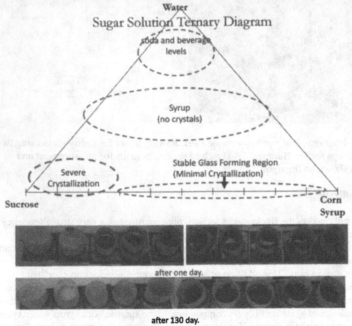

Figure 6. A pseudo-ternary diagram for the sucrose/corn syrup/water system at room temperature, illustrating the strong tendency to crystallize on the high sucrose side. There is no glass formation at room temperature until water is reduced to about 1-2 % by wt. Diagram is a cartoon and not to actual scale. Ratios on the horizontal axis are only approximately linked to the photographs. Below are photos of representative mixtures poured into paper molds with a black bottom to highlight the white crystals.

4.1 Quantitative Analysis of Crystallization

During the making of sugar glass, the experimenter (and cook) becomes painfully aware of both the tendency of the melt to crystallize as well as the importance of avoiding such crystallization in making clear, high quality hard candies. The simple glass system exhibits both moisture-mediated, surface crystallization at room temperature as well as thermally induced crystallization within the bulk, if heated sufficiently above the glass softening point. Sugar glass provides a rich opportunity for the student to explore quantitative experiments on both of these phenomena using relatively simple tools. We have developed experiments to explore both of these aspects with simple home-built apparatus. Our website includes tutorial material to guide the students with their understanding of fundamentals of crystal nucleation and growth.

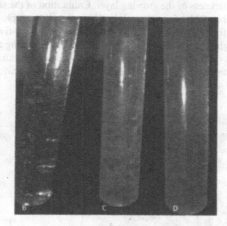

Figure 7. (A) Growth of crystals at the surface (top) of sugar glass in test tube at room temperature after several weeks of high humidity. The photo on the right shows the progression of bulk crystallization from an unheated clear candy glass (B), to one that is heated in a home oven set to 250° F (120°C) for approximately 20 minutes (C & D).

4.2 Effect of Humidity on Crystallization

Initial observations with high sucrose glasses, as well as our standard 2:1 glass, suggested a humidity dependence of the surface crystallization at room temperature. Motivated by the interest of a local high school student in doing a science project, we developed a simple approach for controlling humidity and for measuring crystallization rate. The samples for this study were prepared from circular globs of molten candy pressed between two glass slides and maintained at a uniform thickness by using two steel washers as spacers. Utilizing this sample geometry, only the outer edges of the circular discs of candy glass are exposed to ambient moisture. Thus all humidity-induced, surface crystallization occurs at the outer edge of the sample. The geometry also made it easy to examine and measure the progression of the crystal layer as it grows with time. Simple humidity controlled chambers were constructed from glass cookie jars with tight fitting rubber gasket seals. Three levels of humidity control were achieved using $CaSO_4$ desiccant for the dry chamber, water soaked paper towel for the high humidity chamber and saturated solution of $Mg(NO_3)_2$ for the intermediate (approximate 50%) humidity chamber [24]. Additional details of the method can be found on our website together with a copy of the student's final presentation of her experiments at the regional science competition (PA Jr. Academy of Science).

The results of this experiment were very instructive. Essentially no crystallization occurred on the sample maintained in desiccant, while samples from the 50% relative humidity (RH) chamber grew uniform thickness crystal layer, ideal for measurement. A standard digital camera (in macro mode) was used to record the layer thickness every few days. From a printout of the photograph the student could then use a simple scale (ruler) to measure the thickness of the growing layer. Calibration of the scale was achieved by also measuring the width of the glass slide (1") in the photograph.

The data in figure 8 show the measurements over a twenty-five day period for two different candy glass preparations (143° and 145°C final cooking temperature) including some replication (of the 145° preparation). All of the samples exhibited good repeatability, with crystal growth front (width) that grew at a uniform rate of approximately 0.4 mm per day.

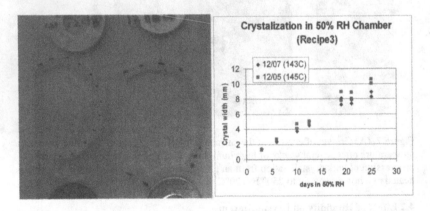

Figure 8. Photograph of crystal growth on outer edge of two samples taken on day 5 after placing in the 50% RH chamber. Note the uniform thickness of the crystal layer in both samples. On right, data show increasing width of crystal layer with time for a standard recipe glass. Two different batches are included (diamonds and squares). Note that the 145° C sample (squares) also includes a replicate.

Samples from the high humidity chamber (essentially 100% RH), did not exhibit the same uniform crystal layer at the outer edge as the 50% RH samples. Instead the outermost layer was clear and liquid in appearance with a thinner crystalline region at the inner growth front, suggesting a subsequent dissolution of the crystals with even higher moisture absorption. This later result was not expected and provides a good illustration of the wealth of interesting behavior for the investigator to discover in this simple system. From these simple experiments many additional experiments for examination come to mind, including the effect of composition.

4.3 Effect of Temperature on Glass Devitrification

Understanding the instability to crystallization underlying both the glassy state and the supersaturated solution from which it arises is an essential aspect of glass science. In glass crystallization is largely precluded by the enormous viscosity of the bulk. However, above the glass transition, the supercooled liquid becomes far more prone to crystallization as the viscosity falls rapidly with temperature, making it easier for the molecules to rearrange into the lower energy crystalline state. We have already discussed how moisture can mediate surface crystallization at room temperature, presumably by providing a lower viscosity path to rearrangement into the crystalline phase. Likewise, as the temperature is increased above the glass transition point, crystallization within the bulk is again possible.

The temperature dependence of crystallization rate is of fundamental importance and a central topic in material science. Sugar glass provides a very easy and accessible gateway to explore the temperature dependent transition from metastable liquid to crystalline state as the viscosity barrier is reduced with elevated temperature.

In order to investigate this temperature dependence, the first experimental obstacle is to achieve appropriate temperature control within the student budget. The home oven, while most convenient, is inadequate for such experiments as it is typically not designed for stable control at these low temperatures and has considerable hysteresis (over and under shoot). For students with access to a laboratory oven with good temperature control, we encourage this approach. However, in the spirit of empowering the student inquiry, we designed a low cost, home-built temperature controlled oven appropriate for this experiment. To minimize cost, a standard light bulb is used as the heat source for an aluminum base plate on which the sample is placed. Heat from the light bulb is controlled by a dimmer switch making low-cost temperature control possible. The temperature of the base plate is monitored by either a low cost digital cooking thermometer or a thermocouple probe. A Pyrex Petri dish provides the cover for the heating plates and the glass sample. A sketch of our apparatus including a plywood base plate and a #10 coffee can to hold the heating plates is shown in figure 9.

Details for the construction are included in our materials. The student is advised to characterize his/her oven by determining temperature vs. time at a range of voltages (e.g. every 10 volts), establishing both the heating time as well as a temp vs. voltage operating chart. This data allows the student to achieve the desired temperature more quickly with a little human assisted voltage ramping. With a little care the student can achieve a controlled temperature in less than about 10 minutes and a temperature uniformity of approximately ± 1°C is typical for this apparatus.

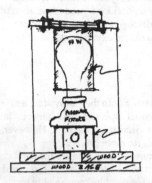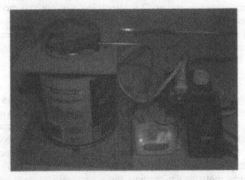

Figure 9. Sketch of the light bulb sample oven for the crystallization vs. temperature studies. (Right) Photograph of the complete apparatus with dimmer control, digital thermometer and digital volt meter (DVM) for monitoring voltage across light bulb heating element.

The samples used in this temperature experiment were the same type of samples described in previous section, i.e. circular glass glob placed between two glass slides (or a glass slide with a cover slit). Here spacers are very important since the material will become fluid at the temperatures required to initiate crystal growth, and without the spacers would ooze out of the glass slides. A calibration of temperature for the sample on the plates can be achieved by using an appropriate low melting standard such as stearic acid.

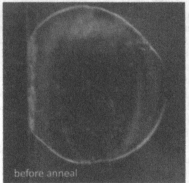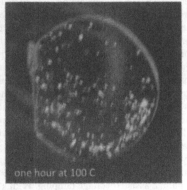

Figure 10. Photograph of crystal growth (right) in a sugar glass sample after one hour at 100°C. The sample on the left is the identical sample prior to heating. Photos taken with low cost Digital Blue computer microscope.

Once the oven temperature is stabilized, the sugar glass sample on microscope slide is placed on the aluminum heating plate of the oven, covered with the Petri dish and allowed to remain in the oven for fixed period of time (nominally 30 minutes). After this fixed period of thermal treatment, the samples are removed and allowed to cool back into the glassy state before examining. Temperatures at 10° increments between 80° and 140°C were found to span the range of interest for our sugar glass. With a T_g of less than 60°C, our material shows no appreciable crystal formation below 80°. Near 150°C (the final cooking temperature), the material begins to boil off water, so this provides an upper limit. Once the sample has cooled, the student examines the sample under crossed polarizers to observe the extent to which crystallization of sugar has occurred. We use a low-cost, computer based digital microscope, the Digital Blue QX5 (available online for under $100, http://www.digiblue.com/). This allows the student to both observe and record the nature and amount of crystallization that has occurred during the thermal treatment period. Any available laboratory microscope with transmission mode and low magnification objectives can be used instead of our Digital Blue. Simply place crossed polarizing sheets above and below the sample. The ability to take a digital photograph for subsequent comparison and analysis is very valuable in this experiment. When no such setup is available, we have found that good results can also be obtained using a standard hand held digital camera placed against the eyepiece of the microscope, while using the digital zoom to achieve a suitable image.

The extent of crystallization can be measured by estimating the area crystallized by eye or, preferably, by utilizing digital image analysis software, such as Image J, an easy to learn, public domain image processing and analysis software developed and distributed by National Institute of Health [25].

In this experiment we are primarily measuring the crystal growth rate, more conveniently expressed as a fraction of area crystallized. Observations of crystal growth with time in our samples show that the initially observed nucleation sites remain constant and that essentially all subsequent growth occurs at these original sites. This experiment provides a wonderful opportunity for the students to consider the distinction between nucleation and growth mechanisms and even come up with their own ideas on how they could establish which process dominates at a given condition. While one cannot see the initial critical nuclei emerge as they are only a few tens of atoms in size, one can often see the consequences of nucleation. These and other very early stage growth experiments in the sugar glass are also possible, but require a higher power microscope and are not part of our discussion here. Nonetheless, this experiment provides much opportunity to observe and ponder new questions and experiments on this topic.

The crystal growth vs. temperature experiments described in this section were carried out by a high school student for her science fair project using the simple method of spot counting for estimating the crystallized area. The experiment was also replicated by a college level REU student who included image processing (with Image J freeware) to measure the fractional area of crystals. In both experiments, the students observed a distinct maximum in the crystal growth rate near 120°C. Example data are shown in figure 11. Additional details for both of the student experiments are included on our website together with their presentations. We have also included a tutorial on nucleation and growth, discussing the topic within the framework of a student with only a high school background in chemistry. There the distinctions between homogeneous and heterogeneous nucleation are discussed as well as basic models for understanding the temperature dependence.

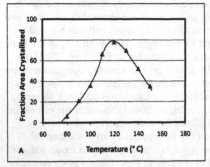 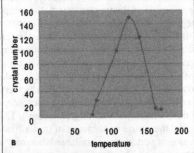

Figure 11. Examples of data measured in two different student experiments. Crystal growth rate vs. temperature in A) done by a college level REU student using Image J to extract the area from photos while B) performed by a high school student using the simple method of estimating the number of "crystal spots". Note the distinct maximum near 120° C for both approaches.

5. Thermal Analysis

Any exploration of the glass state would be incomplete without some consideration of the glass transition phenomena. In a glass research lab today, DSC (differential scanning calorimetry) would likely be the most common method for measuring this property. However, such apparatus is expensive and generally unavailable outside of the research laboratory. Differential thermal analysis (DTA) is somewhat simpler to implement and provides a more straightforward understanding with essentially the same information. We have developed a simple DTA apparatus which can be constructed from items available in most high school laboratories. It consists of measuring the temperature difference between two test tubes, one tube containing the candy glass sample and the other filled with a reference material, while both test tubes are heated in an oil bath. The oil bath is simply a beaker full of vegetable oil placed on a laboratory hot plate equipped with a magnetic stirring bar. The test tubes are held in the oil bath by a wooden holder, which the student can construct, and thermocouples are used to measure the bath temperature as well as the differential temperature between the two tubes while the bath temperature increases. Since exact calibration is not important for the differential temperature (ΔT) measurement, we have constructed the differential pair from a single piece of constantan wire soldered at each end to two pieces of thin copper wire (#24 gauge used, available from telephone hookup wire). A sketch and photograph of our apparatus are shown in figure 13 and additional details on the construction are available on our website.

For the DTA we have chosen to introduce thermocouples as the preferred temperature sensor for three reasons. First, they enable a differential measurement essential to the method; second, they introduce the student to yet another method of temperature measurement used in the laboratory; and third, they allow the extension of this experiment to higher temperatures than the thermistor based digital probes can tolerate.

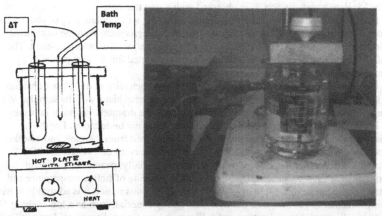

Figure 12. Sketch of the arrangement of a student-built DTA apparatus, together with the photo of a simple implementation with digital thermometer for bath temperature, and a single thermocouple meter for measuring the differential temperature.

In order to obtain a well defined T_g, it is important to quench the sugar glass prior to the measurement run; otherwise, you may not obtain a glass transition. Standard thermocouple meters are adequate to collect the data point by point, and low cost meters are available from about $30 (Harbor Freight). However, to minimize the tedium of such manual data collection, we also included automatic data logging to our DTA experiment. A simple approach would be to use one of the commercial or educational instrumentation (e.g. Pasco, Vernier) [26] available in high school and college labs. However, to remain consistent with our low cost and home-built approach, we also developed our own custom-built data collection instrument utilizing a relatively low cost microprocessor platform. The Basic Stamp Microprocessor (by Parallax, Inc.) was chosen for its low cost, ease of learning and wide popularity within both the educational and hobbyist communities. (There are many sensors designed to interface directly with the Basic Stamp platform, including a thermocouple module (based on DS2760 one wire interface chip) and a humidity sensor chip. There is also a substantial amount of educational material and support available from the Parallax website (http://www.parallax.com/). Although it adds some additional complexity up front, introducing the Basic Stamp for data collection allows the student to learn something about microprocessors, providing a much more flexible, easily adaptable tool for a variety of other experiments involving measurement and control. However, the choice between a commercial vs. home built approach to data measurement and collection will depend on the resources and skills available to the student investigator. Either way, automating the data collection, especially of temperature in thermal scans, enables the student to focus more on the observations and less on the tedium of collecting multiple data points. A more complete description of the details of our apparatus, including the electronic instrumentation, is included on our website.

Sample data for our sugar glass (standard 2:1 recipe) using our DTA are shown in figure 13. Here a very distinct T_g is observed with a step that commences just above 29°C that flattens by 55°C. Using a midpoint definition for T_g, we calculate a T_g of about 42°C for our standard sugar glass recipe. This value is in fair agreement with measurements made on a commercial DSC (TA 2920), where a midpoint of 38°C was measured. The small difference between these two values can be accounted for, at least partially, by differences in thermal history, heating rate, etc.

One should be aware that the details of the thermal behavior in the glass transition can be quite dependent on heating rate as well as the thermal history of the sample. We have had many frustrating opportunities to experience the dramatic variation in results that is possible when proper control of thermal history is not understood. For the measurements reported here, the sample was first quenched from the melt into a beaker of ice water and then allowed to stabilize overnight. Observing this variation of DTA scans with different sample history, while frustrating initially, provides a powerful opportunity for the student to really experience the meaning of fictive temperature and the truly non-equilibrium nature of glassy materials. For this reason it is advisable to have the experimenter begin first with a simple, non-glassy material, such as a low melting point crystalline solid to understand and calibrate the apparatus. Stearic acid was utilized to provide such a reference material. It provides a clear melting transition at about 70°C (and was used as our calibration standard).

The DTA provides the student with a relatively simple, hands-on, yet quantitative, access to the glass transition phenomena. Besides the advantage of low-cost access, our home-built DTA provides the student with a much deeper understanding of the underlying science and instrumentation inherent in the method. Likewise, the transparency of the design permits the student to actually peek and poke at the material as it goes through the various stages of the transformation. One can see the closure of cracks developed on cooling, the softening and subsequent melting of the glassy material and finally the evolution of water vapor bubbles as the material approaches its original preparation (boiling) temperature. Such visual access provides a wonderful stimulus for pondering many aspects related to glass science, indeed materials science.

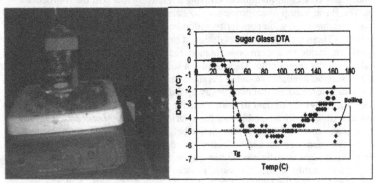

Figure 13. Photograph of the DTA apparatus with Basic Stamp microprocessor for temperature measurement and data logging. A sample DTA scan for the candy glass is shown on the right, with midpoint T_g indicated.

IV. SUMMARY

We have developed a hands-on, mini curriculum for exploring many aspects of glass science through experiments with sugar glasses. Combining a range of activities from material synthesis to measuring basic material properties with student-built equipment, we have developed an informal educational resource designed to facilitate interest and active prolonged engagement of the student with glassy materials in both a relevant and quantitative context. By utilizing a website to house and distribute our resources, we have established a means to reach a large student and teacher population while providing a fast and flexible means to revise and add new material as it is developed.

A complete description of the experiments presented here can be accessed at our website by any interested student or teacher. We include a broad range of materials in a variety of forms (video discussions, experimental procedures, tutorial modules, student presentations, laboratory resources, etc.). The information can be discussed and understood at varying levels of detail, including specific construction methods and procedures, not appropriate in a more formal publication format. Using a web-based approach for distribution of information also allows its regular updating. It is our intention that many additional hands-on activities and experiments will be added on an ongoing basis in the future. Hopefully, this resource will find application in the high school as well as college laboratories, and provide a resource for students and science enthusiasts to explore on their own in their home. We welcome any interesting and appropriate submissions from others in the glass community.

ACKNOWLEDGMENTS

We would like to acknowledge several students who have contributed to testing and developing many of the experiments described here. The summer REU students include Tara Schneider (2006), Sean Kelly (2007) and Sarah Horst (2009). Jung Hyun (Gloria) Noh's interest in pursuing a high school science project in glass inspired two of our experiments on crystallization and Isha Jain's science project (2001) initiated our first exploits into candy glass. Finally, we are most grateful to the NSF for its support through International Materials Institute for New Functionality in Glass (IMI-NFG) as grants DMR-0409588 and DMR-0844014.

REFERENCES

1. National Council for Excellence in Education, *A Nation at Risk: The Imperative For Educational Reform,* (U.S. Department of Education, Washington, DC, 1983)
2. Philip Bell, Bruce Lewenstein, Andrew W. Shouse, and Michael A. Feder, Editors, *Learning Science in Informal Environments: People, Places, and Pursuits* (National Academies Press, Washington, DC, 2009).
3. H. Jain and I. H. Jain, Standard Experiments in Engineering Materials, Science and

Technology, NEW: Update 2000. NASA/CP-2001-211029, pp. 169-182 (2001).

4. H. Jain and I. H. Jain, Proc. Am. Soc. Eng. Edu., Session 3264, pp 1-7 (2002).

5. Harold McGee, *On Food and Cooking: The Science and Lore of the Kitchen* (Scribner, New York, 2004).

6. Richard W. Hartel, *Crystallization in Foods* (Aspen Publishers, Inc., Gaithersburg, MD, 2001).

7. Mathlouthi, M. and P. Reiser (eds), *Sucrose Properties and Applications* (Blackie Academic & Professional, Glasgow, UK, 1995).

8. Bubnik, Z and P. Kadlec, in *Sucrose Properties and Applications,* edited by M. Mathlouthi and P. Reiser (Blackie Academic & Professional, Glasgow, UK, 1995) pp. 107-109.

9. P. Reiser, G. G. Birch and M. Mathlouthi , in *Sucrose Properties and Applications,* edited by M. Mathlouthi and P. Reiser (Blackie Academic & Professional, Glasgow, UK, 1995) p. 190.

10. Hartel, op. cit., p. 271.

11. M. D. Ranken, C. Baker and R. C. Kill (editors), *Food Industries Manual, 24th Ed.* (Blackie Academic & Professional, London, 1997) p. 418.

12. Hartel, op. cit., p. 182.

13. Andrew Van Hook and Floro Frulla, Ind. Eng. Chem., **44**, 1305 (1952).

14. Hartel, op. cit., p. 182.

15. McGee, op. cit., p. 687.

16. Hartel, op. cit., p. 270.

17. McGee, op. cit., p. 682.

18. Louisiana Sugar Planters' Association, *The Louisiana Planter and Sugar Manufacturer*, **46** (American Cane Growers' Association, New Orleans, 1911) p. 87.

19. McGee, p. 681

20. H. Jain and I. Jain, Sugar glass fiber drawing video, [online]. Available from: http://www.lehigh.edu/imi/libraryglassedu.html. Accessed 2009 Nov. 12.

21. John I. Thornton, C. Langhauser and D. Kahane, Journal of Forensic Sciences **29**, 711 (1984).

22. C. Harvey Palmer, *Optics: Experiments and Demonstrations* (The John Hopkins Press, Baltimore, 1962), p. 9.

23. Arthur Hardy and Fred Perrin, *The Principles of Optics* (McGraw-Hill, New York, 1932) p. 548.

24. From Constant Humidity Table in *Handbook of Chemistry and Physics 57th ed.* (The CRC Press, Inc., Cleveland, Ohio, 1976) p. E-46.

25. ImageJ is a public domain, Java-based image processing program developed at the National Institutes of Health. It is available for free download at http://rsbweb.nih.gov/ij/

26. Vernier (http://www.vernier.com/) and Pasco (http://www.pasco.com/) both produce data collection instruments commonly found in high school and college labs.

Mater. Res. Soc. Symp. Proc. Vol. 1233 © 2010 Materials Research Society   1233-PP09-02

Implementation of a Curriculum Leading to a Baccalaureate Degree in Nanoscale Science

R.J. Matyi and R.E. Geer
College of Nanoscale Science and Engineering
SUNY – University at Albany
Albany, NY 12203 U.S.A.

ABSTRACT

The College of Nanoscale Science and Engineering (CNSE) at the University at Albany has developed an academic curriculum leading to the degree of Bachelor of Science in Nanoscale Science. This curriculum represents a 132-credit program designed for completion in eight academic semesters and is consistent with the SUNY General Education Program requirements as implemented at the University at Albany. This curriculum comprises a cutting-edge, inherently interdisciplinary, academic program centered on scholarly excellence, educational quality, and technical and pedagogical innovation. The blueprint for this curriculum is comprised of four basic components: a *"Foundational Principles"* component, a *"Core Competency"* component, a *"Concentration"* component and a *"Capstone Research/Design"* component. The first two components are designed to integrate the dissemination of fundamental, cross-disciplinary, nanoscale science and engineering principles with the cultivation of the critical skill set necessary for advanced undergraduate coursework and interdisciplinary research. The remaining two components expand on these foundational skills to develop the topical expertise, technical depth, and independent research abilities that are essential to a well-rounded undergraduate educational experience. The combination of these instructional tools ensures a customizable and coherent undergraduate degree program that trains the student's intellect how to explore, discover, and innovate, while ensuring its proficiency in a specific nanoscale discipline. The outcome is a unique undergraduate experience that taps into CNSE's global academic leadership in nanoscale science and engineering to attract and educate a diverse and talented pool of qualified scientists and engineers at the baccalaureate level.

INTRODUCTION

Nanoscale science and engineering – the confluence of physics, chemistry, materials science, biology, and engineering to manipulate, measure and control matter at nanometer length scales – represents a revolution in science and technology. At the nanoscale, the behavior of materials and systems is dominated by quantum phenomena and a host of other properties and processes. While unanimity is rarely achieved on any stage, the belief that nanotechnology will revolutionize human life is, by most measures, overwhelmingly accepted by the scientific community.

At the University at Albany (UAlbany), the College of Nanoscale Science and Engineering (CNSE) has been at the forefront of education and research into nanoscale science and technology. The combination of an innovative graduate curriculum, a wide array of cutting-edge research programs, and a unique combination of academic and industrial collaborations located within facilities that define "state of the art", has resulted in international recognition for the CNSE and for UAlbany. The overarching goal of the CNSE, however, is to prepare students to be leaders in this field, since we believe that an adequately trained scientific workforce is essential for creating and transforming the industries that will realize the benefits of nanotechnology [1]. As the nation's first college wholly dedicated to nanoscale science and technology, the CNSE has evolved into an internationally recognized center for research, education, and outreach in nanoscale science, nanoscale engineering, and nanotechnology. The state-of-the-art nanotechnology facilities at CNSE provide students with unique hands-on training and educational experiences that are not currently available at other academic institutions.

An undergraduate curriculum has been designed to extend and reinforce this evolution as well as to further the academic mission of the University at Albany and the State University of New York (SUNY). The curriculum represents a 132-credit program designed for completion in eight academic semesters and is consistent with the SUNY General Education Program requirements, as implemented at UAlbany. It comprises a cutting-edge, inherently interdisciplinary, academic program centered on scholarly excellence, educational quality, and technical and pedagogical innovation. The outcome is a unique undergraduate experience that taps into CNSE's global academic leadership in nanoscale science and engineering to attract and educate a diverse and talented pool of qualified scientists and engineers at the baccalaureate level.

The development of this novel curriculum has been accomplished within significant boundary conditions such as:

- Avoiding an educational experience that is "a kilometer wide but a nanometer deep"
- Providing a foundation in the basic sciences: physical, chemical, biological – also mathematics
- Allowing students to develop competency in their chosen area in nanoscale science
- Providing students with an intellectual foundation that will enable them to become lifelong learners
- Ensuring an interdisciplinary learning experience
- Leveraging the unique capabilities of the CNSE
- Preparing graduating students for a range of post-graduation options (workforce versus advanced academic degrees)

This nanoscale science curriculum has been designed so that it will impart to its students the broad-based (basic and applied) scientific understanding of atomic scale phenomena, behaviors, and properties of matter that is necessary to achieve deliberate control over nanometer-scale atomic and molecular architectures and systems. The program will also enable a quantitative mastery of the fundamental nature of nanoscale interactions, one that can be effectively used to characterize and measure the behavior and structure of nanometer scale assemblies and systems. This degree program as a whole will offer an academically rigorous preparation for students intending to pursue scientific, technical, or professional careers in nanotechnology enabled fields or graduate studies in nanoscale science or nanoscale engineering, as well as other physical sciences such as materials science, physics, and chemistry.

A key aspect of this curriculum is that it has been developed "from the ground up", to maximize academic coherence in nanotechnology. This is contrast with the addition of nano-related courses to pre-existing curricula based in conventional academic disciplines, which appears to be the most common approach to introduce "nano" to established curricula. The fact that concepts in nanotechnology will be used as the vehicle to deliver fundamental concepts in physical, chemical and biological sciences within a comprehensive curriculum is believed to be a unique pedagogical approach.

PRINCIPAL COMPONENTS OF THE CURRICULUM

The CNSE curriculum in nanoscale science is designed to provide undergraduate students with a well-rounded education of the highest quality--one that endows the student's intellect with the analytical tools necessary to explore, discover, and innovate; while cementing the student's basic proficiency and fundamental knowledge in the science of nanotechnology. Table I presents

Year	Fall	Credits	Spring	Credits
1	NSCI 110 – Chem. Principles of Nanoscale Sci and Eng I	4	NSCI 112 – Chem. Principles of Nanoscale Sci and Eng II	4
	AMAT 112 or 118T – Calculus I	4	AMAT 113 or 119T – Calculus II	4
	NSCI 120 – Phys. Principles of Nanoscale Sci and Eng I[1]	4	NSCI 122 – Phys. Principles of Nanoscale Sci and Eng II	4
	NSCI 101 – Nanotechnology Survey[1]	3	NSCI 102 – Societal Impacts of Nanotechnology	3
	GE (optional, may be made up in future semesters)	3	GE (optional, may be made up in future semesters)	3
		15-18		15-18
2	NSCI 124 or 124T – Phys. Principles of Nanoscale Sci & Eng III	4	NSCI 230 or 230T – Thermo. and Stat. Mech. of Nanoscale Systems	3
	NSCI 20x – Science and Eng. Skills elective	2	AMAT 220 – Linear Algebra	3
	AMAT 214 or 214T – Calculus of Several Variables	4	NSCI 220 or 220T – Structure of Matter	3
	GE or liberal arts and sciences elective	3	NSCI 20x – Science and Tech. Skills elective	2
	GE or liberal arts and sciences elective	3	NSCI 20x – Science and Tech. Skills elective	2
			GE or liberal arts and sciences elective	3
		16		16
3	AMAT 314 – Analysis for Applications I	3	NSCI 3XX – Technical Concentration Course	3
	NSCI 3XX – Technical Concentration Course	3	NSCI 360 – Nanoscale Molecular Mat'ls and Soft Matter	3
	NSCI 350 – Intro to Quantum Theory for Nanoscale Systems	3	NSCI 305 – Integrated NanoLaboratory II	3
	NSCI 300 – Integrated NanoLaboratory I	3	NSCI 390 – Capstone Research I: Intro and Lit. Review	3
	GE or liberal arts and sciences elective	3	GE or liberal arts and sciences elective	3
	GE or liberal arts and sciences elective	3	GE or liberal arts and science elective (if needed)	3
		18		15-18
4	NSCI 410 – Quantum Origins of Material Behavior	3	NSCI 4XX – Technical Concentration Course	3
	NSCI 4XX – Technical Concentration Course	3	NSCI 4XX – Elective	3
	NSCI 4XX – Elective	3	NSCI 4XX – Elective	3
	NSCI 490 – Capstone Research II: Team Research and Project Review	3	NSCI 492 or 493 – Capstone Research III: Team Research and Final Report II[5,6]	3
	NSCI 498 – Seminar	1	GE or liberal arts and sciences elective	3
	GE or liberal arts and sciences elective	3	GE or liberal arts and science elective (if needed)	3
		16		15-18

Table I. B.S. in Nanoscale Science Semester-by-Semester Major Academic Pathway (MAP)

the undergraduate curriculum Major Academic Pathway (MAP) that leads to the B.S. degree in Nanoscale Science. Specific components of the curriculum are discussed below.

Core curriculum in nanoscale science

Building on the innovation and success of CNSE's graduate programs in nanoscale science and nanoscale engineering, the core of the undergraduate academic program is comprised of four building blocks designed to preserve both the inherent flexibility required for a true interdisciplinary undergraduate degree and the academic rigor and scholarly excellence demanded by the field of nanoscale science.

1. *Foundational Principles of Nanoscale Science*: The course and laboratory contents of the Foundational Principles component (shown in Table II, with course number, title, and number of credits) are designed to provide the core nanoscale science principles and intellectual "skill sets" required to ensure elementary understanding and basic knowledge of the disciplines of nanoscale science. Two two-semester (Chemical Principles of Nanoscale Science and Engineering I and II; Biological Principles of Nanoscale Science and Engineering I and II) and one three-semester (Physical Principles of Nanoscale Science and Engineering I, II and III) Foundations Principles courses are available, with the choice of "mix" between physical, chemical and biological science dependent in part on the student's planned Nanoscale Science Concentration area. A key feature of all of the Foundational Principles courses is that they will use concepts in nanoscale science and technology to as a portal to the fundamental scientific concepts that are covered in these classes. As discussed below, all Foundational Principles classes will have lab

Foundational Principles of Nanoscale Science		
NSCI 110	Chemical Principles of Nanoscale Science and Engineering I	4
NSCI 112	Chemical Principles of Nanoscale Science and Engineering II	4
NSCI 120	Physical Principles of Nanoscale Science and Engineering I	4
NSCI 122	Physical Principles of Nanoscale Science and Engineering II	4
NSCI 124	Physical Principles of Nanoscale Science and Engineering III	4
NSCI 124T	Physical Principles of Nanoscale Science and Engineering III (Honors)	4
NSCI 130	Biological Principles of Nanoscale Science and Engineering I	4
NSCI 132	Biological Principles of Nanoscale Science and Engineering II	4

Table II. Foundational Principles courses for the B.S. in Nanoscale Science

Core Competencies for Nanoscale Science		
NSCI 220	Structure of Matter	3
NSCI 220T	Structure of Matter (Honors)	3
NSCI 230	Thermodynamics & Statistical Mechanics for Nanoscale Systems	3
NSCI 230T	Thermodynamics & Stat. Mechanics for Nanoscale Systems (Honors)	3
NSCI 300	Integrated NanoLaboratory I	3
NSCI 305	Integrated NanoLaboratory II	3
NSCI 350	Introduction to Quantum Theory for Nanoscale Systems	3
NSCI 360	Nanoscale Molecular Materials and Soft Matter	3
NSCI 410	Quantum Origins of Material Behavior	3

Table III. Core Competencies courses for the B.S. in Nanoscale Science

components that further relate concepts at the macroscale to those at the nanoscale.

2. *Core Competencies for Nanoscale Science*: The course and laboratory contents of the Core Competency component (shown in Table III) is intended to impart the sophisticated capabilities required for advanced, in-depth, study in nanoscale science. Key concepts in the structure and property of solids (both "hard" and "soft") and a significant immersion into quantum theory, thermodynamics, and statistical mechanics will provide the necessary framework for student understanding of nanoscale phenomena. In addition, comprehensive integrated laboratory courses will ensure that students develop the necessary experimental skills that are critical to nanotechnology.

3. *Technical Concentrations in Nanoscale Science*: This component (shown in Table IV) is comprised of specialized advanced undergraduate coursework or individually-directed independent study in a given nanoscale science concentration area. Combined with upper level elective courses this component of the degree permits a high degree of interdisciplinary customization. Currently there are three Technical Concentration areas: (a) nanoelectronics; (b) nanostructured materials, and (c) nanobioscience. These areas represent subfields where the CNSE has significant strength (both in infrastructure and both research and academic capabilities), and the specific Technical Concentration areas are those that best leverage these strengths. In time and with changing faculty research activities and student interests, additional Technical Concentration areas may be added.

NSCI Technical Concentration Courses		
Nanoelectronics		
NSCI 310	Nanoscale Surfaces and Interfaces	3
NSCI 320	Advanced Physical/Chemical Concepts for Nanoscale Science	3
NSCI 420	Electronic Properties of Nanomaterials	3
NSCI 421	Nanoscale Electronic Devices	3
NSCI 422	Concepts in Molecular Electronics	3
NSCI 423	Magnetic and Spintronic Materials and Devices	3
NSCI 424	Optoelectronic Materials and Devices	3
Nanostructured Materials		
NSCI 310	Nanoscale Surfaces and Interfaces	3
NSCI 320	Advanced Physical/Chemical Concepts for Nanoscale Science	3
NSCI 430	Nanoscale Physical Properties in Reduced Dimensions	3
NSCI 431	Growth of Nanostructured Materials	3
NSCI 432	Particle Induced Chemistry	3
NSCI 433	Properties of Nanoscale Composite Structures	3
NSCI 434	Nanostructural Characterization Techniques	3
Nanobioscience		
NSCI 240	Biochemical Principles for Nanoscale Science	3
NSCI 330	Energetics and Kinetics in Nanobiological Systems	3
NSCI 440	Biological Architectures for Nanotechnology Applications	3
NSCI 441	Nanobiology for Nanotechnology Applications	3
NSCI 442	Nanoscale Bio-Inorganic Interfaces	3
NSCI 443	Biological Routes for Nanomaterials Synthesis	3

Table IV. Technical Concentration courses for the B.S. in Nanoscale Science

Capstone Undergraduate Research in Nanoscale Science		
NSCI 390	Capstone Research I: Introduction and Literature Review	3
NSCI 490	Capstone Research II: Team Research and Project Review	3
NSCI 492	Capstone Research III: Team Research and Final Report	3
NSCI493	Capstone Research III: Team Research and Final Report (Honors)	3

Table V. Capstone Research/Design courses for the B.S. in Nanoscale Science

4. *Capstone Undergraduate Research/Design in Nanoscale Science*: This component (shown in Table V) entails 3 semesters of individually-directed independent research that will serve as an on-site internship within academic/industrial research and design teams. Individually-directed independent research will serve as an on-site internship within a true research environment that is conducive to innovation and discovery. Initial plans are to utilize the state-of-the-art facilities in the CNSE for the undergraduate capstone research experience. In this way, potential capstone research projects will draw on the rich environment of company partners on site in the CNSE and have the potential to expose students to real-life problems. This will provide enrichment to students who are pursuing either industrial careers or advanced degree options

Taken as a whole, the four components merge and integrate basic and advanced course and laboratory work with customized skills training and individually-directed independent research. This combination of pedagogical tools ensures a customizable and coherent undergraduate degree program and teaches the student how to explore, discover, and innovate, in addition to being well proficient in a specific nanoscale discipline.

Supporting elements to the nanoscale science curriculum

In addition to the technical courses in nanoscale science discussed above, students following the path towards the B.S. degree in Nanoscale Science will be advised to also take a variety of courses in mathematics, as well as Nanotechnology Survey Courses and various science and technology skills electives.

- *Mathematics*: The importance of a solid foundation in basic and applied mathematics is self-evident; the curriculum ensures that students will have adequate preparation though courses including multivariate calculus, statistics, linear algebra, vector analysis, Fourier series, and ordinary differential equations.

- *Nanoscale Science and Technology Skills electives*: State-of-the-art research, development, and deployment in nanotechnology are, in many cases, facilitated by a skill- and tool-set of advanced experimental and computational capabilities. As a result, we will offer courses (shown in Table VI) in areas such as instrument control, engineering design, and modeling that will ensure that graduating students have hands-on capabilities in these critical focused technical areas.

- *Nanotechnology Survey Courses*: A series of Nanotechnology Survey courses (shown in Table VII) will be offered to program participants and UAlbany students, at large, to

NSCI Science and Technology Skills Courses		
NSCI 201	Computer Control of Instrumentation	2
NSCI 202	Introduction to Nanoscale Engineering Design and Manufacturing	2
NSCI 203	Advanced Circuits Laboratory	2
NSCI 204	Finite Element Modeling	2
NSCI 205	Numerical Simulation	2

Table VI. Science and Technology Skills courses for the B.S. in Nanoscale Science

NSCI Nanotechnology Survey Courses		
NSCI 101	Nanotechnology Survey	3
NSCI 102	Societal Impacts of Nanotechnology	3
NSCI 103	Economic Impacts of Nanotechnology	3
NSCI 104	Disruptive Nanotechnologies	3

Table VII. Nanotechnology Survey courses for the B.S. in Nanoscale Science

introduce critical concepts and approaches in nanoscale science and nanoscale engineering and stimulate undergraduate discourse in related topics including societal, economic, and cultural impacts of nanotechnology.

General Education requirement

The General Education ("GenEd") Program at the University at Albany proposes a set of knowledge areas, perspectives, and competencies considered by the University to be central to the intellectual development of every undergraduate. GenEd coursework is intended to provide students with a foundation that both prepares them for continued work within their chosen major and minor fields and gives them the intellectual habits that will enable them to become lifelong learners. Courses within the program are designed not only to enhance students' knowledge, but to provide them as well with new ways of thinking and with the ability to engage in critical analysis and creative activity. The General Education Program at the University at Albany consists of a minimum of 30 credits of coursework in the following areas: disciplinary perspectives, cultural and historical perspectives, and communication and reasoning competencies. Our curriculum in Nanoscale Science integrates the GenEd requirements into the overall 132 credit requirement for graduation.

LEARNING OUTCOMES FOR THE NANOSCALE SCIENCE PROGRAM

Numerous learning outcomes for this curriculum have been designed to ensure that students demonstrate the technical and professional proficiencies necessary to enable the identification, description, discovery, experimental investigation, and theoretical interpretation of nanoscale phenomenon and, as a result, become highly successful scientists, educators, and leaders in the global "innovation economy" of the 21st century. The anticipated learning outcomes for students who graduate from this degree program are shown in Table VIII.

147

Learning Objectives for the Nanoscale Science B.S. Degree Curriculum	
1	Graduates will exhibit basic knowledge of mathematics, particularly statistics, linear algebra, multivariate calculus, and differential equations, and the foundational principles of nanoscale science necessary for knowledge development and theoretical and experimental problem-solving in nanoscale systems and architectures.
2	Graduates will possess the analytical abilities and scientific know-how to systematically design, conduct, and complete successful experimental procedures and applied tasks, including the skill set to analyze and interpret data from a variety of sources (experiment, simulation, etc).
3	Graduates will have the ability to analyze, construct or conceive a theoretical or practical nanoscale system, component, or process necessary to elucidate the fundamental nanoscale properties of a given atomic level phenomenon or architecture to meet desired target specifications within realistic and practical constraints. The latter are those that would typically be encountered in the real world of applied nanoscale research and development.
4	Graduates will be able to comprehend the cross-disciplinary nature of a nanoscale problem (technical and/or non-technical aspects) when appropriate and, as a result, be capable of functioning and contributing as successful team members within inter-disciplinary R&D organizations.
5	Graduates will have the knowledge to identify, analyze, deconstruct, and solve well-defined, open-ended, or poorly-defined research problems in nanoscale science.
6	Graduates will have a clear understanding of their professional, societal, and ethical responsibilities as scientists, researchers, educators, and responsible members of society at large.
7	Graduates will demonstrate clear and effective communication skills, both oral and written, in technical and non-technical environments.
8	Graduates will have the broad "well-rounded" education profile necessary to understand, appreciate, and direct the impact of nanoscale science phenomena, innovations, and applications at the local, regional, and global economic, environmental, technological, and societal levels. This comprehensive education will be achieved through an appropriate integration of the UAlbany General Education requirements with the nanoscale science curriculum requisites.
9	Graduates will acquire an embedded recognition of the need for, and responsive ability to engage in life-long, self-motivated learning.
10	Graduates will exhibit pertinent and timely knowledge of contemporary issues, discoveries, and events that are related to nanoscale science, nanoscale engineering and nanotechnology and their potential impact on industry, business, and society.
11	Graduates will possess the techniques, skills, and modern engineering tools necessary for effective nanoscale science research.

Table VIII. B.S. in Nanoscale Science Learning Outcomes

DISCUSSION

The CNSE undergraduate curriculum is intended to attract and retain the large numbers of qualified undergraduate students who are presently inaccessible to SUNY and the private institutions of higher learning in New York State. This inaccessibility is driven by the lack of the interdisciplinary nanoscale science and nanoscale engineering degrees that are sought by this rapidly growing sector of the university clientele, as documented almost invariably by each study, blueprint, report, and analysis published by every governmental body, corporate organization, academic entity, think tank, and cross-organizational panel across the globe-- including the National Science Foundation, which forecasts the need for more than two million nanotechnology educated professionals at all employment levels in the U.S. by 2014, with another five million nanotechnology jobs being required worldwide in support fields and disciplines. As such, the curricula proposed offer unique educational opportunities that are designed to create a highly qualified pool of future scientists, engineers, researchers, and educators in the emerging fields of nanoscale science and nanoscale engineering, while helping train and retrain pertinent sectors of the high technology workforce in New York State.

Thus there is ample justification for the creation of an undergraduate curriculum in nanoscale science. However, the fact remains that the development and implementation of such a curriculum has been accompanied by a host of unique issues (both problems and opportunities) that influenced the curriculum development activity. Some of these issues are addressed below.

The interdisciplinary nature of nanotechnology education in an academic structure

One of the hallmarks of nanotechnology is that it is an inherently interdisciplinary and cross-disciplinary field: concepts of materials science, physics, chemistry, biology, engineering, and other academic disciplines are necessary for the study, manipulation and fabrication of materials and structures at the nanoscale. In many academic settings, the process of bringing together a wide array of academic fields within a traditional department-centered university environment can be challenging.

In contrast, the CNSE was established from the ground-up without a departmental structure. Unencumbered by boundaries between traditional disciplines, CNSE students are able take part in a groundbreaking interdisciplinary course of study designed to address the needs and challenges of nanoscale research, rather than simply putting nano-labels on established science. The result is a unique, cross-cutting education environment that establishes the fundamental concepts necessary for understanding nanoscale phenomena, and also offers the opportunity to delve more deeply into a wide variety of advanced topics needed to support cutting-edge research.

Breadth versus depth versus time to graduate

While a highly interdisciplinary undergraduate educational experience has *prima facie* desirability, it has an inherent problem of breadth versus depth. As mentioned earlier (and in the language of nanotechnology), an explicit boundary condition invoked during the development of this curriculum was to avoid an educational experience that is "a kilometer wide but a nanometer deep". Ideally, then, an interdisciplinary undergraduate experience – one that unites the chemical, physical, and biological concepts that underpin much of modern nanotechnology – would include roughly equal classroom and laboratory exposure of these fundamental concepts.

Unfortunately, the serious constraint of constructing a program that was consistent with our learning objectives while permitting a matriculating student to graduate after eight academic semesters (coupled with other constraints, such as the New York state mandated General Education requirements) made it difficult to design a program that would provide (a) equal coverage in chemistry, physics and biology fundamentals, (b) sufficient fundamental and applied coursework in nanotechnology, (c) appropriate preparation in mathematics, and (d) compliance with the SUNY General Education requirement. In order to resolve this issue, a solution has been implemented in which students who opt for a Nanobioscience Technical Concentration will be able to substitute classwork in the Biological Principles to be substituted for more the corresponding Physical Principles courses.

Laboratory experience

A successful laboratory experience is an integral component of undergraduate education in the physical and biological sciences. In a nanotechnology curriculum, however, there exists the obvious difficulty that nanoscale phenomena exist at the nanoscale – something that may not be conveniently accessed by conventional undergraduate laboratory instruments. In implementing the new curriculum in nanoscale science, the CNSE has been able to leverage its unique capabilities to address this problem. CNSE's Albany NanoTech Complex is a $5 billion, fully-integrated research, development, prototyping, and educational facility that provides strategic support through outreach, technology acceleration, business incubation, and test-based integration support for onsite corporate partners including IBM, Tokyo Electron, Applied Materials, ASML and International SEMATECH, as well as other "next generation" nanotechnology research activities. As a result, the infrastructure and personnel required to operate and maintain a state-of-the-art nanotechnology laboratory environment pre-date the initiation of the undergraduate program and are available to bring the teaching laboratory on-line.

It is important to note that from its formative stages, the commitment was made to integrate the undergraduate student experience (and in particular, their laboratory classes) with the pre-existing academic/industrial ecosystem that is characteristic of the CNSE. At the beginning of their undergraduate experience, for instance, the students will experience laboratory science through the Foundational Principles classes. Since these introductory classes will be taught using examples in nanotechnology, the labs will follow with extensive use of methods such as atomic force microscopy for nanoscale characterization and chemical vapor deposition of nanoscale structures. This hands-on approach to using nanotechnology as a portal to a discussion of fundamental scientific concepts is without precedent but will be the normal mode of operation in the CNSE curriculum.

The laboratory infrastructure will be dual use, in that it will support both the teaching needs of the College as well as being available for the numerous ongoing research activities in the CNSE. Although significantly more expensive to set up than traditional introductory chemistry or physics teaching labs, this approach will enhance the educational goals and directly support the learning outcomes for our undergraduate population. The integration of the undergraduate laboratory experience is expected to conclude in the capstone research/design project. Here, we fully expect that students will conduct their research programs in the most advanced facilities available at the CNSE, such as NanoFab North: a 228,000 square foot, $175 million facility including 35,000 square feet of cleanroom space with Class 1 capable 300mm wafer production. We believe that by bringing our undergraduate students into this one-of-a-kind R&D facility, we will provide them a laboratory experience that is unmatched in the world.

Industrial relevance

As stated earlier, one of the boundary conditions of the development of the CNSE undergraduate program is to give students the opportunity to develop the skill needed to join the nanotechnology workforce following graduation. Our on-site industrial partnerships have placed the CNSE into a unique position of understanding workforce development issues, and the undergraduate curriculum reflects that knowledge. We have designed our coursework so that students will be fully equipped to become productive workers in nanotechnology and related fields immediately upon graduation (either with a CNSE industrial partner, or elsewhere). Of course, should students choose to move on for post-graduate academic studies, we believe that the curriculum will provide an excellent basis for continued work towards either the M.S. or Ph.D. degrees.

Assessment

Assessment – the continuous quality improvement process for higher education – is a systematic process in which faculty define goals and objectives for programmatic activities, develop performance metrics to gauge success, and use evaluation results to retool what is done in and out of the classroom so that who serve as teachers can get better and better at what they do. The University at Albany has had a tradition of embracing the goals of academic assessment, both due to its intrinsic value in improving the quality of our teaching, but also because of its growing importance of demonstrating to the various constituents we serve that we are responsible stewards of the public investment made by the citizens of our state.

The CNSE recognizes that the initiation of an entirely new curriculum in an area such as nanoscale science deserves (if not requires) a systematic, broad-based, and multi-pronged approach to assess student's progress towards and achievement of the learning outcomes. One reason is needs-driven – many aspects of this curriculum in nanoscale science are quite novel, and it will be imperative to establish early on which aspects are successful at achieving the learning objectives, and which ones are not. Beyond satisfying this obvious requirement, however, the establishment of a wholly-new undergraduate curriculum offers the unique opportunity to fully incorporate assessment activities into the learning environment from the start. At the least, three primary metrics will be employed: course-embedded assessment; student-peer assessment; and capstone experience-driven assessment. In addition, enhanced assessment approaches are being explored to determine if more fundamental information regarding teaching and learning in nanotechnology education can be gained from an extended analysis of the undergraduate program.

CONCLUSIONS

The baccalaureate curriculum in nanoscale science exploits the unparalleled academic, professional, and infrastructural resources of the College of Nanoscale Science and Engineering and its $5 billion Albany NanoTech research and development and education complex. By leveraging CNSE's one-of-a-kind physical infrastructure, interdisciplinary faculty, and extensive public-private partnerships, the proposed undergraduate curriculum will hold a scholarly profile and pedagogical impact that are singularly distinct from and highly complementary to current academic offerings at the remaining SUNY campuses and other New York State institutions of higher learning. The curriculum will also serve as an effective tool in the attraction of the highest quality undergraduate students from around the world to UAlbany.

REFERENCES

1. "Nanotechnology: Societal Implications—Maximizing Benefits for Humanity: A Report of the National Nanotechnology Initiative Workshop" (eds M. C. Roco and W. S. Bainbridge, December 2003)

Mater. Res. Soc. Symp. Proc. Vol. 1233 © 2010 Materials Research Society   1233-PP10-05

Evolution of the Women in Materials Program: A Collaboration Between Simmons College and the Cornell Center for Materials Research

Velda Goldberg[1], Leonard J. Soltzberg[1], Michael D. Kaplan[1], Richard W. Gurney[1], Nancy E. Lee[1], George G. Malliaras[2], and Helene R. Schember[3]

[1]Physics and Chemistry Departments, Simmons College, Boston, MA 02115, U.S.A

[2]Materials Science and Engineering Department, Cornell University, Ithaca, NY 14853, U.S.A.; current address: Centre Microélectronique de Provence, Ecole Nationale Supérieure des Mines de Saint Etienne, 880 route de Mimet, 13541 Gardanne, FRANCE

[3]Cornell Center for a Sustainable Future, Cornell University, Ithaca, NY 14853, U.S.A.

ABSTRACT

The *Women in Materials* (WIM) program is an on-going collaboration between Simmons College and the Cornell Center for Materials Research (CCMR). Beginning in 2001, during the initial four years of the project, materials-related curricula were developed, a new joint research project was begun, and nearly 1/2 of Simmons College science majors participated in materials-related research during their first two years as undergraduates. We have previously reported the student outcomes as a result of this initial stage of the project, demonstrating a successful partnership between a primarily undergraduate women's college and a federally funded Materials Research Science and Engineering Center. Here, we report the evolution and impact of this project over the last three years, subsequent to the initial seed funding from the National Science Foundation. The *Women in Materials* project is now a key feature of the undergraduate science program at Simmons College and has developed into an organizing structure for materials-related research at the College. Initially, three faculty members were involved and now eight faculty members from all three laboratory science departments participate (biology, chemistry, and physics). The program now involves research related to optoelectronics, polymer synthesis, biomaterials, and green chemistry, and each semester about 80% of the students who participate in these projects are 1st and 2nd year science majors. This structure has led to enhanced funding within the sciences, shared instrumentation facilities, a new minor in materials science, and a spirit of collaboration among science faculty and departments. It has also spawned a new, innovative curricular initiative, the *Undergraduate Laboratory Renaissance,* now in its second year of implementation, involving all three laboratory science departments in incorporating actual, on-going research projects into introductory and intermediate science laboratories. Most importantly, the *Women in Materials* program has embedded materials-related research into our science curriculum and has deepened and broadened the educational experience for our students; the student outcomes speak to the program's success. Approximately 70% of our science majors go on to graduate school within two years of completing their undergraduate degree. Our students also have a high acceptance rate at highly competitive summer research programs, such as Research Experience for Undergraduates (REU) programs funded by the National Science Foundation.

INTRODUCTION

There is considerable evidence that research participation motivates undergraduate science students and strengthens their preparation for graduate study or for the workplace.[1,2,3,4,5] An innovative collaboration between a research university (Cornell University) and a predominantly undergraduate institution (Simmons College) has underscored the validity of this idea and has spawned a comprehensive redesign of the laboratory program at Simmons based on research participation in place of traditional closed-end laboratory experiments.

THE PROGRAMS
Women in Materials

The *Women in Materials* (WIM) program, begun in 2001, is based on a research collaboration between Simmons College physics and chemistry faculty and members of the Cornell Center for Materials Research (CCMR) and has been focused on investigating factors that lead to the degradation of organic light emitting diodes (OLEDs). The details of this program have been described elsewhere.[6] As shown below in the OUTCOMES section, a large number of Simmons students have benefited from research participation under the aegis of the WIM program; at any given time the numbers of WIM participants constitute about 25% of our chemistry and physics majors, with about 80% of these students starting in WIM research as 1st or 2nd year students. Students participating in WIM research have jointly authored, along with Simmons and CCMR faculty, work presented at national Materials Research Society and American Chemical Society meetings and published in refereed journals. The WIM program has led to an increase in the number of women majoring in physics, minoring in materials science, enrolling in materials-related NSF-REU programs, and entering Ph.D. programs in materials science or related disciplines.

Undergraduate Laboratory Renaissance

The demonstrated success of the WIM program has led us to integrate ongoing research participation into the very fabric of our laboratory science curricula. In the *Undergraduate Laboratory Renaissance* (ULR) program, we are replacing traditional closed-end ("expository") laboratory experiments with research participation in most chemistry courses and in selected physics and biology courses. This approach to laboratory education includes some special innovations such as using a Wiki as a communal laboratory notebook to coordinate the work of multiple lab sections in a course; and employing junior and senior undergraduates working on a research project for their senior research as "post-docs" to help coordinate and supervise work in the lower-division lab sections.

The formal launch of the ULR in 2008-09 exceeded our expectations. The Summer 2008 faculty training/planning sessions and the academic year implementation of research integration produced insights into organization and operation of research-integrated laboratory sections, publishable research results, and research synergies. We developed a schema for formalizing the goals of existing laboratory experiments so that they can be incorporated into ULR labs without loss of student learning. Six courses enrolling about 200 students have now been brought online with research-integrated laboratories: Organic Chemistry I, Organic Chemistry II, Quantitative Analysis, Biochemistry, Molecular Biology, and Microbiology of Food, Water and Waste.

Assessments of student attitudes (see under OUTCOMES) revealed an enthusiastic response to being involved in actual research, although a small number of students found the responsibility of research participation somewhat daunting. Students cited the value of being responsible for selecting materials and methods and became highly invested in the projects, often coming to lab outside of their scheduled sections to check on progress.

In addition to attitudes, we assessed achievement of laboratory skills to be sure that students in this program continue to learn the concepts and skills expected from the laboratory program. We found that students in ULR labs did as well or somewhat better on these assessments compared with students in the previous year's non-ULR labs (see under OUTCOMES).

OUTCOMES

Table 1. Student Participation in the WIM Program (1st – 3rd Year Students)

Academic Year	Number of Student Participants
2001-2002	29
2002-2003	28
2003-2004	38
2004-2005	30
2005-2006	24
2006-2007	23
2007-2008	23
2008-2009	41

Table 2. Evaluation of Student Attitudes in the ULR Program

Please evaluate the Team, the Wiki, and the research based laboratory experience, using a scale of 1 – 5.
5 = Strongly Agree   1 = Strongly Disagree

Team:

	Statistics
1. The Team functioned as a single unit.	3.55 ± 1.1
2. I did my part to insure that my Team functioned as a Team.	4.36 ± 0.6
3. The sum of the Team's effort was greater than each part.	3.84 ± 1.1
4. I did what I agreed to do in preparation for the laboratory.	4.52 ± 0.5
5. The division of labor during the lab was equitable. (All work was equally shared by all team members.)	3.31 ± 1.2
6. I lived up to my own expectations for this project.	4.31 ± 0.7

Wiki:

	Statistics
7. The Wiki facilitated the group planning, organization and completion of the Pre-lab work.	3.88 ± 1.0
8. The Wiki facilitated the group reporting and sharing of data and results during and after the laboratory.	3.82 ± 1.0
9. The Wiki facilitated the group completion of a final report.	3.63 ± 1.1
10. The Wiki enabled me to learn from other lab sections.	3.53 ± 1.1
11. The Wiki hindered the five-week research experience, and I would have preferred to have used a lab notebook.	2.54 ± 1.3
12. I would have preferred to write individual lab reports, because the work was not equally shared among individuals.	3.20 ± 1.4
13. I would suggest the similar use of Wiki's for a research based laboratory experience in the future.	3.97 ± 1.0

As a result of the research based laboratory experience:

	Statistics
14. I believe I have learned something valuable that will help me in future laboratory courses.	4.64 ± 0.7
15. my appreciation of a well-written, well-structured lab report has increased.	4.35 ± 0.7
16. my understanding of all necessary components of a thoughtful, complete lab report (such as procedure and discussion) has increased.	4.18 ± 0.8
17. I think I have learned something that will help me write a better lab report.	4.15 ± 0.9
18. my appreciation of completing Pre-Lab work or (preparing before attending a lab) has increased.	4.42 ± 0.7
19. I have gained insights into Scientific Research that I think will stick with me for the rest of my life.	4.33 ± 0.8
20. I am more confident in my laboratory skills.	4.51 ± 0.6
21. I am more confident in my ability to operate the IR instrument.	4.61 ± 0.6
22. I am more confident in my ability to interpret IR spectra.	4.41± 0.7
23. I better appreciate how to apply and use IR spectra to determine the outcome of a reaction.	4.44 ± 0.7
24. at some point during the semester, I was personally encouraged or considered: pursuing a career in research, changing my major to chemistry / biochemistry, pursuing a minor in chemistry / biochemistry.	3.39 ± 1.5 2.63 ± 1.4 3.09 ± 1.6
25. at some point during the semester, I was more confident in my decision to continue my pursuit of a career in research, of a major in chemistry / biochemistry, of a minor in chemistry / biochemistry.	3.28 ± 1.6 2.70 ± 1.5 2.98 ± 1.6

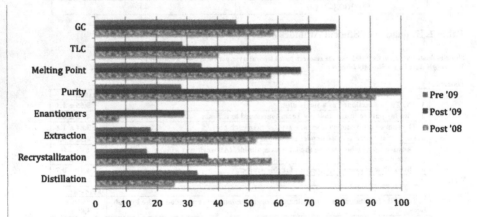

A written assessment of laboratory skills for Organic Chemistry I was administered to a cohort of students the summer after they completed the standard expository laboratory curriculum ("Post-Lab '08" textured bars). The identical written assessment was administered to a second cohort before ("Pre-Lab '09" grey bars) and after ("Post-Lab '09" black bars) the redesigned, research-integrated laboratory curriculum. (Students repeating the course were eliminated from the 2009 statistics).

Figure 1. Sample Assessment of Achievement in the ULR Program: Pre- and Post-laboratory Exam in Organic Chemistry I

Insights from the First Year ULR Implementation

- the WIKI-based electronic notebook works as intended but requires careful faculty monitoring;
- for courses with multiple lab sections, a one hour per week common pre-lab is very desirable;
- the upperclass Research Mentors function as intended, capably serving in the role played by post-docs in a research university setting;
- students' acquired laboratory skills are as good or better than those achieved in non-ULR labs;
- the goal of publishable data emanating from the combined work of multiple laboratory sections in a course is a realized benefit of the ULR approach;
- the goal of underclass students in ULR labs wanting to continue the research in the following year is a realized benefit of this approach.

DISCUSSION

Research participation for undergraduate science students brings notable benefits to both students and faculty. It is desirable and feasible to bring students into the research laboratory as early as the freshman year. Integration of ongoing faculty research into the curriculum-based laboratory program requires considerable analysis of educational goals, careful planning of the lab organization, and on-the-fly adaptation during the semester. The outcomes in student enthusiasm and achievement are well worth this investment. There is also the potential to leverage faculty time available for research in an undergraduate institution by increasing the number of students working on a project.

ACKNOWLEDGMENTS

We are grateful for support from the National Science Foundation, Directorate for Mathematical & Physical Sciences, Divisions of Materials Research and Chemistry under grant numbers DMR-0108497 and DMR-0605621; and from the W.M. Keck Foundation.

REFERENCES

1. "Research is Another Word for Education" R. Wilson, A. Cramer, and J.L Smith, UCLA, Los Angeles, CA: in "Reinvigorating the Undergraduate Experience" L. Kauffman and J. Stocks, eds., Council on Undergraduate Research 2004.
2. "Integrating Technology, Science and Undergraduate Education at Smith College: the Creation of Student-Faculty Research Centers" S.P. Scordilis and T.S. Litwin, Clark Science Center, Smith College, Northampton, MA: CUR Quarterly March 2005.
3. "Spinning Straw Into Gold" M.A. D'Agostino, Department of Biology, Franklin and Marshall College, Lancaster, PA: CUR Quarterly June 2004: Creating Time for Research Vignettes.
4. "Organic Chemistry Lab as a Research Experience" T.R. Ruttledge, Department of Chemistry, Earlham College, Richmond, IN 47374: *J. Chem. Educ.* **1998** *75*, 1575-1577.
5. "A New Project-Based Lab for Undergraduate Environmental and Analytical Chemistry" G. Adami, Department of Chemical Sciences, University of Trieste, Trieste, Italy: *J. Chem. Educ.* **2006** *83* 253-256.
6. "Women in materials: a collaborative effort between Simmons College and the Cornell Center for Materials Research," Velda Goldberg, George G. Malliaras, Helene Schember, Michael

Kaplan, Leonard Soltzberg, Richard W. Gurney, Patrick Johnson, in *Forum on Materials Science Education,* edited by Trevor R. Finlayson, Fiona M. Goodchild, M. Grant Norton, Scott R. J. Oliver (*Mater. Res. Soc. Symp. Proc.* **909E**, Warrendale, PA, 2006), 0909-PP03-10.

AUTHOR INDEX

SUBJECT INDEX